EMPower
extending mathematical power

Seeking Patterns, Building Rules

Algebraic Thinking

TEACHER BOOK

TERC
Because math and science build futures

Mary Jane Schmitt, Myriam Steinback, Tricia Donovan, and Martha Merson

TERC 2067 Massachusetts Avenue
Cambridge, Massachusetts 02140

EMPower Authors / Research and Development Team
Mary Jane Schmitt, Myriam Steinback, Tricia Donovan and Martha Merson

Contributors and Reviewers
Sarah Adams, Donna Curry, Marlene Kliman, Pam Meader, Heidi Schuler, and Sherry Soares

Technical Team
Production and Design Team: Juania Ashley, Valerie Martin, and Sherry Soares
Copyeditor: Jill Pellarin

Evaluation Team
Brett Consulting Group:
Belle Brett
Marilyn Matzko

EMPower™ was developed at TERC in Cambridge, Massachusetts. This material is based upon work supported by the National Science Foundation under award number ESI-9911410 and by the Education Research Collaborative at TERC. Any opinions, findings, and conclusions or recommendations expressed in this publication are those of the authors and do not necessarily reflect the views of the National Science Foundation.

TERC is a nonprofit made up of teams of math and science education and research experts dedicated to innovation and creative problem solving. At the frontier of theory and practice, TERC's work encompasses research, content and curriculum development, technology innovation, professional development, and program evaluation. TERC has a passion for social justice and strives to create level playing fields for all learners, reaching more than three million students every year.

https://www.terc.edu/empower

Printed in the United States of America

1 2 3 4 5 6 7 8 9 10

ISBN 978-1-7373997-2-8

Send all *EMPower* content inquiries to:

Adult Numeracy Center at TERC
empower@terc.edu

Adult Numeracy Center

Advisory Board

Donna Curry, Equipped for the Future, National Center
Solange Farina, Math Exchange Group, New York City
Linda Kime, Mathematics Association of America and University of Massachusetts, Boston
Ron Kindig, Los Angeles Unified School District (retired)
Kenn Pendleton, General Education Development (GED) Testing Service
Ujwala Samant, National Center for the Study of Adult Learning and Literacy
William Thomas, American Mathematics Association of Two-Year Colleges (AMAYTC) and University of Toledo

Consultants and Reviewers

Scott W. Beckett, Jacksonville State University
Sherry Fraser, Interactive Mathematics Program
Arthur G. Fruhling, Yuba Community College
Nancy R. Johnson, Manatee Community College
Steve Monk, University of Washington
Elizabeth Phillips, Connected Mathematics Program
Mark Rudd, Albuquerque TVI
Karen Sitren, Santa Fe Community College
Cornelia Tierney, TERC
Kevin Wheeler, Three Rivers Community College

EMPower Teacher Participants

Arizona

Gila River Indian Community 21st Century Community Learning Center, Gila River
Andrea Parrella

Illinois

Adult Education Center, Rock Valley College, Rockford
Phyllis Flanagan, Melinda Harrison

Comprehensive Community Solutions, Inc., YouthBuild, Rockford
Ryan Boyce

Township High School District 214 Community Education, Arlington Heights
Kathy Conrad

Massachusetts

Action for Boston Community Development, Inc., LearningWorks, Boston
Kelly Qualman

Adult Learning Center at Middlesex Community College, Bedford
Roberta Froelich

Brockton Adult Learning Center, Brockton
Marilyn Moses

Community Learning Center, Cambridge
Nellie Dedmon, Sylvia Lotspeich Greene, Linda Huntington

Dimock Commuity Health Center, Roxbury
Martha Gray

Harvard Bridge to Learning and Literacy, Cambridge
Judy Hikes, Carol Kolenik

Holyoke Adult Learning Opportunities (HALO) Center, Holyoke
Kelly Martin, Glenn Yarnell

Hyde Square Task Force, Jamaica Plain
Nora O'Connor

Jewish Vocational Services, Boston
Susan Arase, Terry Lerner, Jeff Snyder

Lowell Adult Education Program, Lowell
Barbara Goodridge

Mount Wachusett Community College, Gardner
Patricia Vorfeld

Notre Dame Education Center, South Boston
Esther D. Leonelli

Pioneer Valley Adult Education Center, Northampton
Olu Odusina

Project Place, Boston
Sharon Carey, Maia Hendrickson

Quinsigamond Community College Adult Basic Education Program, Worcester
Cathy Coleman

Read/Write/Now, Springfield
Lee E. Boone, Susanne Campagna, Michelle Faith Brown, Lucille Fandel

Somerville Center for Adult Learning Experiences (SCALE), Somerville
Tom Glannon

X-Cel Adult Education, Dorchester
Don Sands

Maine

Bailey Evening School, Bath
Pam Bessey

Gardiner Adult and Community Education Program
Diann Bailey

Windham Adult Education, Windham
Eva Giles

EMPower Teacher Participants *(continued)*

New Jersey

New Brunswick Public Schools—Adult Learning Center, New Brunswick
Jacqueline Arkoe, Phyllis Boulanger, Sue Helfand, Karen C. Pickering

New York

Adult Learning Center, La Guardia Community College, Long Island City
Norma Andrade, Mark Trushkowsky

BEGIN Managed Programs, New York
Charles Brover, Solange Farina, Santiago Perez, Elizabeth Reddin

Borough of Manhattan Community College/CUNY Adult Basic Education Program, Center for Continuing Education and Workforce Development, New York
Elliot Fink, Mark Lance

City University of New York, Continuing Education, New York
Lisa Simon

Lehman Adult Learning Center, Bronx
Deidre Freeman

SafeSpace-LifeSkills, New York
Jonna Rao

The Adult Learning Center, College of Staten Island, Staten Island
Karen Johnsen

Pennsylvania

Lancaster-Lebanon Intermediate Unit 13/Career Link, Lancaster
Margaret Giordano, Barbara Tyndall

Rhode Island

Dorcas Place, Parent Literacy Center, Providence
Shannon Dolan, Jerelyn Thomas

Tennessee

Hawkins County Adult Education Program, Rogersville
Lisa Mullins

Acknowledgments

Many friends and family members supported the teachers' and *EMPower's* efforts. Thank you Denise Deagan, Judith Diamond, Cara Dimattia, Ellen McDevitt, Sandy Strunk, and the board of the Adult Numeracy Network. Roberta Froelich, Brad Hamilton, Michael Hanish, David Hayes, Judy Hikes, Alisa Izumi, Carol Kolenik, Esther D. Leonelli, Myra Love, Lambrina Mileva, Luz Rivas, Johanna Schmitt, Rachael Stark, Jonathan Steinback, and Sean Sutherland all made unique and timely contributions to the project.

We appreciate the encouragement and advice from John (Spud) Bradley, *EMPower's* program officer, and Gerhard Salinger of the National Science Foundation.

We are indebted to every adult student who participated in the piloting of *EMPower*. Their honest feedback and suggestions for what worked and what did not work were invaluable. We could not have completed the curriculum without them or their teachers.

Contents

Introduction to the EMPower Series

Background

*Extending Mathematical Power (*EMPower*)* integrates recent mathematics education reform into the field of education for adults and out-of-school youth. *EMPower* was designed especially for those students who return for a second chance at education by enrolling in remedial and adult basic education programs, high school equivalency programs, and developmental programs at community colleges. However, the series is appropriate for a variety of other settings as well, such as high schools, workplaces, and parent and paraprofessional education programs. *EMPower* builds interest and competency in mathematical problem solving and communication.

Over the course of four years (2000–2004), a collaboration of teachers and researchers with expertise in adult numeracy education and K–12 mathematics reform developed and piloted eight contextualized units. These units are organized around four central topics: number and operation sense; patterns, functions, and relations; geometry and measurement; and data and graphs. The *EMPower* program serves as a model for a cohesive mathematics series that offers content consistent with the *Principles and Standards for School Mathematics* (NCTM, 2000), as well as frameworks that are adult-focused, such as the *Equipped for the Future Content Standards* (Stein, 2000), the *Massachusetts ABE Curriculum Frameworks for Mathematics and Numeracy* (Massachusetts Department of Education, 2001), and the *Adult Numeracy Network's Framework for Adult Numeracy Standards* (Curry, Schmitt, & Waldron, 1995). The program fosters a pedagogy of learning for understanding; it embeds teacher support and is transformative, yet realistic, for multilevel classrooms.

EMPower challenges students and teachers to consistently extend their ideas of what it means to do math. The series focuses on mathematical reasoning, communication, and problem solving with a variety of approaches and strategies, not just rote memorization and symbol manipulation. The program fosters a learning community in which students are encouraged to expand their understanding of mathematics through open-ended investigations, working collaboratively, sharing ideas, and discovering multiple ways for solving problems. The goal of *EMPower* is to help people build experience managing the mathematical demands of various life situations, such as finances and commerce, interpretation of news stories, and leisure activities, and to connect those experiences to mathematical principles.

A Focus on Mathematical Content

The *EMPower* series supports students' and teachers' growth by directing attention to significant mathematical understandings.

EMPower emphasizes:

- Data analysis, geometry and measurement, algebra, and number and operations at all student levels.
- Reliance on benchmark numbers—such as powers and multiples of 10, common fractions, and their decimal and percent equivalents—for making mental calculations.
- Early use of calculators to support computation.
- Development of reasoning on proportion and parts of quantities before consideration of formal operations with rational numbers.
- Making decisions about data where students generate, as well as interpret, graphical representations.
- Geometry and measurement based on opportunities to see and touch in developing an understanding of spatial relationships and formulas.
- Leading with patterns and relationships in contextual situations and the representations of these situations with diagrams, tables, graphs, verbal rules, and symbolic notation to develop algebraic competence.

A Focus on Pedagogy

Mathematics is meaningful within a social context. While mathematical truths are universal, the meaning and relevance of numbers changes according to the setting and culture. Therefore, the *EMPower* pedagogy is focused on sets of connected activities that require communication and discourse.

EMPower asks students to

- Work collaboratively with others on open-ended investigations;
- Share strategies orally and in writing; and
- Justify answers in multiple ways.

Key features of activities provide

- Clear mathematical goals
- Contexts that are engaging and useful for young people and adults
- Opportunities to strengthen mathematical language and communication skills
- Various ways of entering and solving problems
- Puzzles that draw students into problems and motivate them to seek a solution.

Overview of EMPower Units

Features of the Teacher Book

FACILITATING LESSON **1**

Countries in Our Closets

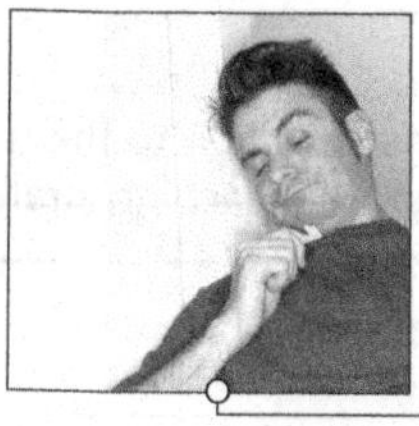

Where were your clothes made?

Synopsis

This is the first of a set of lessons on categorical data in which students collect, examine, sort, and organize their own data, and then compare their data to another data set.

1. Students bring to the class data on the different geographic locations where their clothes were made.
2. The class combines its data into one frequency graph, and students make several statements about it.
3. Working in pairs, students compare their class graph to data from another class.
4. Pairs of students each create a new frequency graph, organizing the class data by continent. Students notice which continent is most prevalent.
5. Discussion centers on comparing country and continent graphs, closing with the question How does the story change when the graph's categories change?

Objectives

- Use a frequency graph to organize data
- Identify the story that the data tell
- Compare data from various samples
- Change the categories and articulate the change in the story

EMPower™ Facilitating Lesson 1: Countries in Our Closets 15

Each lesson opens with a **synopsis** of the lesson activities.

Objectives summarize the skills students should gain from the lesson.

Materials/Prep

- Grid paper
- Paper strips for posting statements
- Post-it® Notes
- Rulers

As an aid for determining in which continent the countries are located, have on hand a world atlas such as *The New Children's Illustrated Atlas of the World*, by Keith Lye (Philadelphia: Courage Press, 1999). For a listing of countries in Asia, see *Resource: Countries in Asia, Blackline Master 4.*

Make sure you have assigned *Countries in Our Closets* (*Student Book*, p. 7). Collect and bring in about 20 items of clothing in case your students have not collected their own data. (See *Facilitation*, p. 25, for how to handle this.)

Prepare a transparency or make a copy of the graph "Countries in Our Closets, Springfield, MA" (*Student Book*, p. 10).

Opening Discussion

Ask students:

Are clothes made in this city? In this country? Why or why not?

Why would anyone care where his or her clothes were made?

Acknowledge that we should know something about where our clothes come from, given how clothing is worn next to our skin. There are also tremendous economic and political issues at stake as manufacturing jobs are outsourced outside of the United States.

Heads Up!

If students do not bring in data, see *Facilitation*, p. 25.

You have looked in your closets, checked the tags on some of your clothes, and collected information. What did you learn?

Which country do most of the clothes you checked come from?

Post the country names in a place where you can easily refer to them later.

Introduce the terms **data set** and **mode.** Each student's list is a data set, and each data set could have a mode—the country name that occurs most frequently.

Vocabulary note: If students are searching for words to express their thoughts, offer terms and concise definitions. Students can keep track of these in their student books. For example, **data** are the information collected. The data that are collected make up the **sample.** The whole group from which information could be collected is the **population.** See *Lesson 1 Commentary*, p. 24, for more information.

16 Many Points Make a Point: Data and Graphs EMPower™

The **Materials/Prep** feature tells teachers what materials they will need to prepare for the day's activities.

The **Opening Discussion** gives teachers ideas for engaging students at the beginning of the lesson.

Heads Up! alerts teachers to helpful hints and things to look for in facilitating the lesson.

Dialog balloons and bold text suggest statements and questions teachers can use to spark and guide class discussion.

A **Summary Discussion** guides the conclusion of the lesson.

Summary Discussion

Finish the lesson by posing the following questions. (Students may not have ready answers at this point, but the questions will be a thread through the rest of the lessons.)

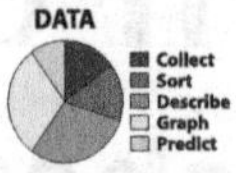

What stories can be told from this set of data (or what questions can be asked and answered)?

How does organizing the information into different categories impact how the data may be interpreted?

Who would be curious about what the people in this class eat?

Direct student to the *Reflections* questions (*Student Book*, p. 145).

Practice, Extension, and **Test Practice** sections list and describe choices for additional work.

Practice

Take a Sample, p. 24
For practice choosing a representative sample.

Thirsty and *Ouch!*, pp. 26 and 28
For practice organizing information and making statements.

Extension

Friends and Family Drink Too, p. 30

Data Collection, p. 31
Students begin recording what they eat for a project they will do in a later lesson. For now, they just write down all the foods they eat throughout the day.

Test Practice

Test Practice, p. 32

Looking Closely draws the teacher's attention to the goals and suggests teaching strategies.

Looking Closely

Observe whether students are able to

Organize data into consistent categories

Do students group similar foods together? Make sure students are moving index cards around and that they are writing down the category names. If students seem stuck, have them ask you specific questions.

If you see foods categorized incorrectly, ask a question such as "Do all group members agree with the categories?" Encourage students to think, question, and defend their reasoning. If students use overlapping categories such as "cereal" and "grain," present an example close to their own experience. Say, for instance, "It is

EMPower™ Facilitating Lesson 2: Most of Us Eat... 33

Math Background helps teachers deepen their understanding of mathematical concepts.

LESSON 1 COMMENTARY

Rationale

The lesson takes students through the first steps of displaying data: collecting and organizing the information. As you work through this lesson, you and your students will notice that each time you categorize and recategorize the data set, you tell a different story.

Math Background

Data are information. This information may be numerical—e.g., salaries, test scores, heights, age, weight—as well as categorical—e.g., countries, foods we eat, types of product defects.

Most data are reported using statistics based on fractions and percents—one-third of pregnant mothers, half of native plants, etc. Commonly used, or benchmark, fractions and percents are often invoked to influence others to make decisions. When the headlines shout, "One-half of all pregnant women …," readers tend to form opinions based on the data. What readers often neglect to consider is the size of the sample, where the data were collected, and how the data were organized.

The media generally report data that are based on a sample. It is rare for any group to have the time or resources to ask every person in the population to contribute information. The idea of sampling is to study a part in order to gain information about the whole. The sample—its size and characteristics—influences the data and the conclusions of the study.

A representative sample by definition includes representation across the population being surveyed. For example, a representative sample of the community would include individuals from across town, not just one neighborhood. Likewise, a representative sample across America would include individuals from many different states and regions, not just one or two. The size of the sample is important when inferences are made. For example, 30 samples could be sufficient for a population of several hundreds. The deciding factor for determining sample size is how confident you want to be about the inference.

It is not the concern of this unit to determine appropriate sample size, but it is important to note that statisticians use formulas derived from repeated surveys to decide sample size. Regardless, small samples are used in many studies. Factors ... might lead ... outcome.

... k critically ... or into ... ield ... esson 2.

EMPower™

The **What to Look For** grid allows teachers to track students' progress and think ahead about next steps.

EMPower™

WHAT TO LOOK FOR IN *LESSON 1*	WHO STANDS OUT? (LIST STUDENTS' INITIALS)			NOTES FOR NEXT STEPS
	STRONG	ADEQUATE	NEEDS WORK	
Concept Development • Records all data • Organizes data • Recognizes key features of frequency graph • Compares data • Changes categories and notes change in story				
Expressive Capacity • Uses precise language to describe the frequency graphs • Contributes to the discussion about the way changing categories affects the story the data tell				
Use of Tools • Uses graph paper • Uses atlas or other reference				
Background Knowledge • Draws on understanding of geography, relative size, and number sense for comparisons				

Facilitating Lesson 1: Countries in Our Closets 23

The authors give ideas for **Making the Lesson Easier** and **Making the Lesson Harder.**

Context

Some students may know about *maquiladoras* in Mexican border towns, where women make clothes for very little money and with no benefits or environmental Occupational Safety and Health Administration (OSHA) workplace protections. CorpWatch (www.corpwatch.org) is one source for information on *maquiladoras.*

Facilitation

If students do not bring in data, or if their sample is too small, skip the second part of the *Opening Discussion.* Have available a pile of 20 clothing articles with labels. First, ask students to predict where the clothes were made. Post the list of their guesses. Note that it will be hard for them to answer this question unless they organize the information on the labels. Then divide up the 20 articles of clothing. Have students write the name of the country for each piece of clothing on a Post-it Note, one country name per note. Ask: "Where are most of our clothes made?" Then continue with the activity.

Making the Lesson Easier

Frequency graphs lend themselves to comparisons among categories. If students have little fluency stating comparisons, you may choose only to compare size, using terms like "greater," "fewest," or "less than." For students who are encountering data formally for the first time, the notion that collapsing data yields different stories may be difficult. Treat this lightly in the activity, and revisit such questions after students have more experience categorizing and recategorizing data in the homework and in *Lesson 2.*

Making the Lesson Harder

If your students can handle benchmark fractions and percents, get them to look critically at the data, including the source and sample size. You might ask:

If we asked another class what countries are in their closets, what do you think would happen to the categories? What if we asked the entire community?

How do you think your data would compare to data from another class of adult students in another community?

If students struggle with the idea of sample, you might try this: Have them each write their favorite color on a Post-it Note. If you have a small class, ask them to write the color on two Post-it Notes. Place all of the notes in a container. Have someone randomly (eyes closed) choose a few notes from the container and place them across a line to form a frequency graph. Ask the students how they think this sample compares to the actual total number of colors on notes in the container. You can have them do another frequency graph to compare the sample to the actual total.

Facilitating Lesson 1: Countries in Our Closets 25

LESSON 1 IN ACTION

Alice articulates the mathematical principle behind compressed data.

I asked, "How did the change in categories affect what we noticed about the data?"

Alice answered, "Well, we keep losing information."

"How so?"

Patiently, Alice explained that when we started our work, every bit of data was visible. She added that we had lost details initially recorded. "At first, we knew every country in every person's closet and how many pieces of clothing came from that country. Then we combined the data, and we lost track of who had which countries. Then we did it by continent, and we lost track of all the countries."

Alice's realization quickly gained agreement from the rest of the class. After all, just the previous week a classmate had noted, "When you change the amount of data you look at, you find different things."

Sonia added her comment with increased conviction: "It is like politics. Politicians use a graph and tell you this is true, but you look at the graph, and it does not tell you everything."

Tricia Donovan
Pioneer Valley Adult Education Center, Northampton, MA

26 Many Points Make a Point: Data and Graphs EMPower™

In **Lesson in Action,** EMPower teachers share their classroom experiences.

Overview of EMPower Units

Features of the Student Book

LESSON **1**

Countries in Our Closets

Where were your clothes made?

In a world where information is at our fingertips, it is important to be able to make sense of it. What does it tell us? Why do we care?

Begin with your own clothes. Where were they made? Find out! You will collect **data** by reading the labels on your clothes. You will organize that data, make statements about it, and note where most of yo were made.

You will then try organizing the data in a different way. Ho the data in new groups change the story of the countries in You will compare your class data to data of other classes. W be the same?

On Your Own

I think the country in which most of my clothes were made

Look at the labels of at least eight items of clothing. Make a track of the information you are collecting. Research questi were your favorite clothes made?

EMPower™ Lesson 1: Countries in C

An **introduction** explains the purpose and content of the lesson.

Vocabulary words are easy to identify in bold type.

Activity 1: Organizing the Data

Look at the data you collected about countries where your clothes were made.

Organize the data.

1. What strikes you about your data?

2. What is the **mode** for your data?

3. What is the mode for the class data?

8 Many Points Make a Point: Data and Graphs EMPower™

The class **activity** comes next. Often students work with a partner or in a small group.

Practice: Categorically Speaking

Sometimes we can organize categories into subgroups of other categories in order to better understand the data. For example, drug stores will often organize their aisles by category—skin care, beauty aids, and so on. Then the items in each of those aisles are also organized. This makes it easier to take inventory and place orders and for the customer to find products.

Can you think of other examples where information is categorized?

Use the chart below *or* create your own chart; draw a picture; make a list; or write about examples you see at home or at work. In the chart below, one example is given.

Location	What Stuff or Information?	How Is It Organized?
Drug store	Over-the-counter products they sell	Skin care Stationery Beauty aids First aid Seasonal Hair care

EMPower™ Lesson 1

Practice pages invite students to use new skills.

Extension: Taking Inventory

Find a drawer or closet filled with many things. Use a frequency graph to show the contents. Start with six categories. Then do another frequency graph to show the same items, but this time use three categories.

For example.

My Medicine Cabinet

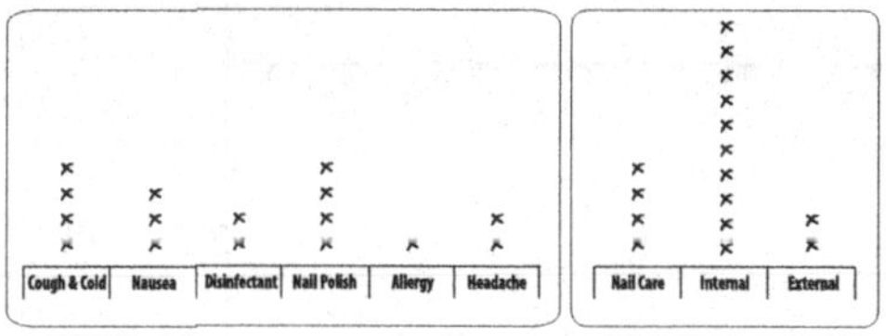

16 Many Points Make a Point: Data and Graphs EMPower™

Extension pages present a challenge or a new but related math idea.

Ample space allows students to record most work in their books.

A **Test Practice** that reflects the format of the GED test concludes each lesson.

Test Practice

1. Freda worked at a printing press. She kept a tally of all the defects she found in the books she was processing. Based on the frequency graph she created, shownbelow, which of the following statements is true?
   (1) There were more defects related to colors than any other defect.
   (2) There were twice as many torn pages as there were warped spines.
   (3) There were more warped spines than colors that bled.
   (4) Half of all the defects were related to colors that bled.
   (5) Half of all the defects were related to pages that were torn.

Book Defects Frequency Graph

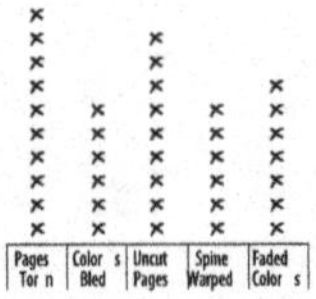

2. Tony, a tour guide, has been keeping a tally of visitors from different states. According to his tally, which of the following is *not* a true statement?
   (1) The fewest number of visitors came from Oregon.
   (2) There were as many visitors from Maine as there were from Georgia.
   (3) There were as many visitors from Georgia as there were from Oregon and Texas.
   (4) There were twice as many visitors from Florida as there were from Rhode Island.
   (5) There were twice as many visitors from Idaho as there were from Maine.

U.S. Tourist Frequency Graph

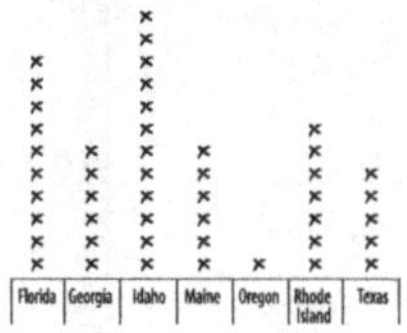

3. Clara is paid a commission on each of the large electronics devices that she sells. She tallied the different items she sold for the month. Based on her tally, what can she tell her boss about her sales?
   (1) She sold twice as many DVD players as she did TVs.
   (2) She sold twice as many DVD players as she did computers.
   (3) She sold more DVD players than she did TVs and computers combined.
   (4) She sold more TVs and VCRs than she did stereos and DVD players combined.
   (5) She sold half as many TVs as she did stereos.

Home Media Frequency Graph

4. Bronson, a forest ranger in the Green Mountains, kept track of the different animals that were reportedly seen in one of the campgrounds during the month of May. Based on his tally, which of the following statements could he tell the media?
   (1) There were twice as many wolves reportedly seen as bears.
   (2) Half of all the reported sightings were wolves.
   (3) There were more bears than wolves reportedly seen.
   (4) There were twice as many moose reportedly seen than there were black bears.
   (5) One-quarter of all reported sightings were brown bears.

North American Wildlife Frequency Graph

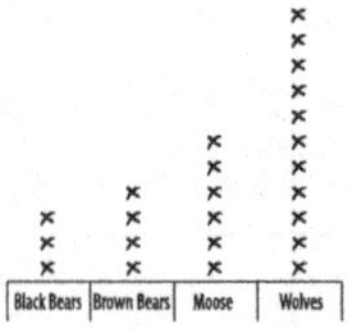

5. According to Global Exchange, some workers in China who make clothing for Disney are paid as little as $0.16 per hour. At this wage, what amount would a worker make for a 40-hour workweek?
   (1) $64.00
   (2) $6.40
   (3) $6.00
   (4) $80.00
   (5) $0.60

6. In 2003 the U.S. Department of Labor (DOL) issued new regulations for overtime pay. The DOL estimated that under the new regulations 1.3 million low-wage workers would become eligible for overtime pay, unless their wages were raised to $425 per week. The DOL estimated that 24.8 percent of those workers were Hispanic and 16.6 percent were African American. What percent were neither Hispanic nor African American?

Changing the Culture

The authors have created this series to follow the National Council of Teachers of Mathematics (NCTM) Principles and Standards; however, every teacher who uses *EMPower* faces the challenge of transforming the prevailing culture of his or her math classroom. *EMPower* pilot teachers offer some ideas for facilitating this transition:

- Set the stage. As a class, set ground rules. Explicitly state that this is a space for everyone to learn. As one teacher said, "We are in this together. Share, even if you do not think you are right. Whatever you add will be helpful. It lets us see how you are looking at things."
- Group your students. Match students whose learning styles and background knowledge complement each other. Ask questions, such as, "How did it go to work together?" "How did everyone contribute?"
- Allow wait time. Studies have shown that teachers often wait less than three seconds before asking another question. Students need more time to think.
- Sit down. Watch students before interrupting to help them. Listen for logic and evidence of understanding. Follow the thread of students' thinking to uncover unconventional approaches. During discussions with the whole group, hand over the chalk.
- Review written work. Look beyond right and wrong answers to learn everything you can about what a student knows. Determine what seems solid and easy, as well as patterns in errors. If students are scattered, suggest ways they can organize their work; this is likely to lead to more efficient problem solving and clearer communication.
- Question. Hearing the right answer is not necessarily a cue to move on. Question students at this point too. Specific questions are included in the lesson facilitation.

Unit Sequences and Connections

The sequence in which the *EMPower* units can be used effectively with your class will depend on the backgrounds and interests of students. The units are not numbered so teachers can order them according to their class needs; however, the authors suggest specific unit arrangements that will support students' progression through certain concepts.

The authors do not recommend sequencing the units according to the traditional basic math model that begins with whole numbers, follows with fractions, decimals, and percents; data and graphs; algebra; and then geometry. Instead, the authors suggest you integrate the five units that focus on numbers with the units on geometry, data, and algebra. The authors found this integration of topics helped to motivate the adult students in their pilot classes.

Although the units were not specifically designed to build on one another, there are clear connections between some of the units in the series. *Over, Around, and Within: Geometry and Measurement* provides a nice introduction to the program because it focuses on small whole numbers. *Everyday Number Sense: Mental Math and Visual Models* could follow to further develop whole number mental math skills and visual models. *Using Benchmarks: Fractions and Operations* provides the necessary groundwork with fractions, decimals, and percents to describe approximate relationships between data sets in *Many Points Make a Point: Data and Graphs.* And *Split It Up: More Fractions, Decimals, and Percents* continues to expand students' repertoire of familiar fractions, decimals, and percents. *Seeking Patterns, Building Rules: Algebraic Thinking* builds upon the tools and relationships used in *Keeping Things in Proportion: Reasoning with Ratios.*

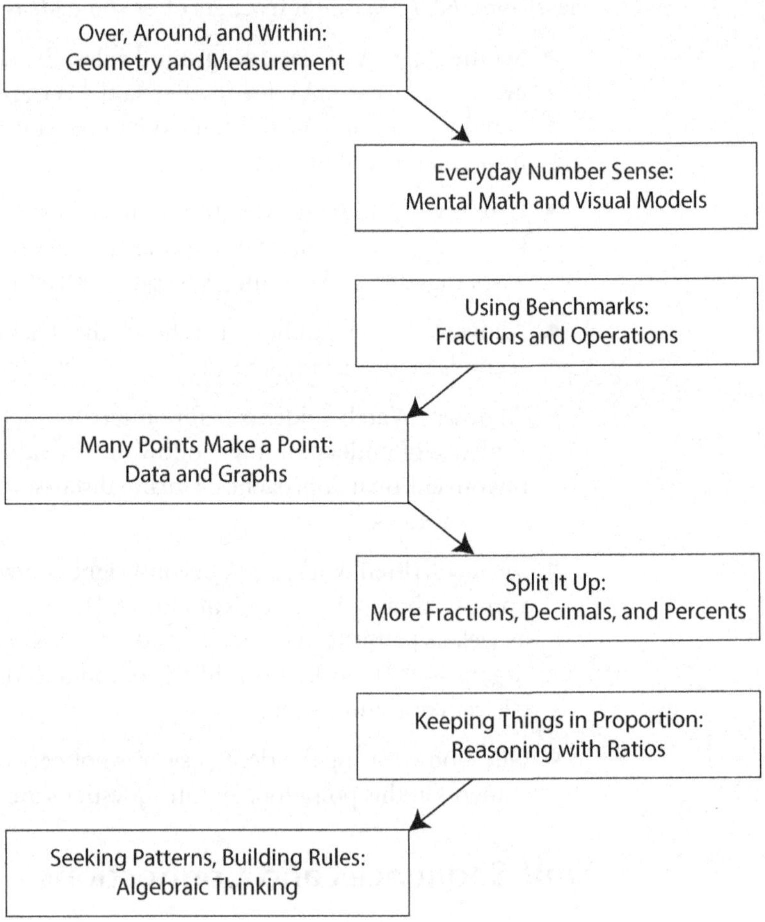

Unit Descriptions

Over, Around, and Within: Geometry and Measurement
Students explore the features and measures of basic shapes. Perimeter and area of two-dimensional shapes and volume of rectangular solids provide the focus.

Everyday Number Sense: Mental Math and Visual Models
Students solve problems and compute with whole numbers using mental math strategies with benchmarks of 1, 10, 100, and 1,000. Number lines, arrays, and diagrams support their conceptual understanding of number relationships and the four operations.

Using Benchmarks: Fractions and Operations
Students use the fractions $\frac{1}{2}$, $\frac{1}{4}$, $\frac{3}{4}$, and $\frac{1}{10}$; the decimals 0.1, 0.5, 0.25, and 0.75; and the percents 50%, 25%, 75%, 100%, and the multiples of 10% as benchmarks to describe and compare all part-whole relationships.

Many Points Make a Point: Data and Graphs
Students collect, organize, and represent data using frequency, bar, and circle graphs. They use line graphs to describe change over time. They use benchmark fractions and

the three measures of central tendency—mode, median, and mean—to describe sets of data.

Split It Up: More Fractions, Decimals, and Percents
Building on their command of common benchmark fractions, students add thirds, eighths, hundredths and their decimal and percent equivalents to their repertoire of part-whole relationships.

Keeping Things in Proportion: Reasoning with Ratios
Students use various tools—objects, diagrams, tables, graphs, and equations—to understand proportional and nonproportional relationships.

Seeking Patterns, Building Rules: Algebraic Thinking
Students use a variety of representational tools—diagrams, words, tables, graphs, and equations—to understand linear patterns and functions. They connect the rate of change with the slope of a line and compare linear with nonlinear relationships. They also gain facility with and comprehension of basic algebraic notation.

Frequently Asked Questions

Q: *I have classes that are widely multilevel. Can this work?*

A: Many teachers see a wide range of levels within the group as an obstacle. Turn the range of levels to your advantage. Focus on students' representations (words, graphs, equations, sketches). This gives everyone the chance to see that answers emerge in several ways. Slowing down deepens understanding and avoids facile responses. Having calculators available can even the playing field. Implement the suggestions in *Making the Lesson Easier* and *Making the Lesson Harder* of each lesson facilitation in the *Lesson Commentary* sections.

Q: *How do I deal with erratic attendance patterns?*

A: Uneven attendance can be disruptive. Students who miss class may feel disoriented; however, the lessons spiral back to the most important concepts. When the series circles back, students will have a chance to revisit concepts and get a toehold.

Q: *What do I do if I run out of time, and there is no way to finish a lesson?*

A: Each activity is important, but reviewing it is equally important. It is better to cut the activity short so there is time to talk with students about what they noticed. Maximize the time by selecting a student or group whose work you feel will add to the class's understanding to report their findings. Be conscious of when you are letting an activity go on too long because the energy is high. Fun is good, but be sure important learning is happening. If you like to give time in class to reviewing homework, and you want to hear from everyone in discussions, you will run out of time. Schedule a catch-up session every three or four lessons.

Q: *How do I respond to comments such as "Can't we go back to the old way?"*

A: Change is unsettling, especially for students who are accustomed to math classes where their job is to work silently on a worksheet solving problems by following a straightforward example. Be clear about the reasons why you have chosen to de-emphasize some of the traditional ways of teaching in favor of this approach. Ultimately, you may need to agree to some changes to accommodate students' input. Meanwhile, stick with the program. Reiterate for students what they have accomplished. When there is an "Aha!" moment, point it out.

Q: *My own math background is not strong. Will I be able to teach this series?*

A: Yes! Most teachers tend to teach the way they were taught. Adopting a different stance requires support, and the more types of support, the better. This series offers support in a few ways. The teacher books for each unit list open-ended questions designed to keep the math on track. In the *Lesson Commentary* sections, *Math Background* helps teachers deepen their understanding of a concept. In addition, the *Lesson in Action* sections provide examples of student work with comments that illuminate the underlying mathematics.

The best support often comes from a colleague. If no one at your site is currently teaching *EMPower*, join the Adult Numeracy Network, http://adultnumeracynetwork.org, and attend your regional NCTM conference. Look for others who are integrating the NCTM Principles and Standards through the use of a curriculum such as *Investigations in Number, Data, and Space* (TERC, 1997); *Connected Mathematics* (Lappan et al., 1998); or *Interactive Mathematics Program* (Alper et al., 1997).

Unit Introduction
Seeking Patterns, Building Rules: Algebraic Thinking

Major Themes

Viewing algebra as a tool to better understand situations

> *Algebra can be seen as a language with "dialects" of literal symbols, graphs, tables, words, diagrams, and other visual displays... To represent a situation in familiar terms is often a gateway to understanding the situation. Thus, algebra is a way of thinking about and representing many situations. It has a language and a syntax, along with tools and procedures, that promote this thinking and modeling (National Research Council, 1998, p. 163).*

Seeking Patterns, Building Rules emphasizes patterns and relationships as well as language and representation to describe various situations. Because the investigations in most lessons of this book begin with a real-life situation, mathematical modeling—or describing a situation with mathematics—is the central activity. With a situation as the starting point, students look for patterns and relationships in order to

- Understand what is changing and how it is changing in a situation;
- Describe the generality of that situation; and
- Predict future outcomes for that situation.

In this unit, students use **multiple representations** to mine the patterns and relationships inherent in situations. Tables, graphs, verbal descriptions, diagrams, and algebraic equations are not ends in themselves; rather, they are tools that help students magnify the meaning and better see what exactly is going on in a situation as well as where it might lead.

This is the unit's icon, which serves to remind students and teachers that the situation is a central focus of the activity.

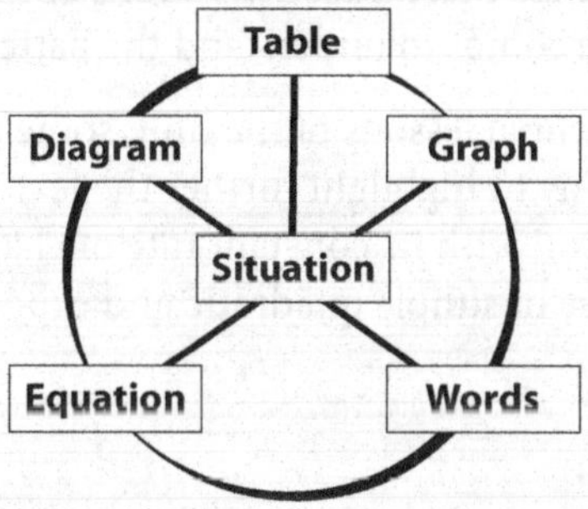

Seeking and describing patterns in order to predict outcomes and generalizations

> *What humans do with the language of mathematics is to describe patterns. Mathematics is an exploratory science that seeks to understand every kind of pattern—patterns that occur in nature, patterns invented by the human mind, and even patterns created by other patterns (Steen, 1990, p .8).*

Whether reading health pamphlets or financial reports or generating statements about local conditions or personal needs, adults need to be able to generalize. Throughout this unit, students become accustomed to asking and answering questions such as

- What comes next?
- What is the relationship between x and y?
- What is changing here? How fast is it changing? What is staying the same?

Multiple representations of patterns and relationships

The process of understanding situations where patterns of data are represented in diagrams, tables, graphs, equations, or words—for example, the break-even point for two situations, the length of time before an investment doubles, or whether a sales offer makes sense given personal circumstances—becomes easier when students have the ability to represent the mathematics of the situation in a variety of ways. Furthermore, as students develop the ability to flexibly represent numerical situations, they are better able to make predictions, comparisons, and decisions because they are better able to generalize.

Links between multiple representations

The ability to make connections between the slope of a line in a graph, the rate at which numbers increase or decrease in a table, and the constants in a formula can deepen a person's understanding of a phenomenon. Students make more sense of the various representations by comparing them. They also develop the ability to imagine or predict what data represented in another form should look like, a skill that allows them to know immediately whether two representations are linked.

Unique characteristics of linear patterns and functions

This unit focuses on one of the most basic types of patterns and functions, the linear function. As students link and compare different representations, they come to understand the characteristics of linear patterns—the constant rate of change, the straight-line graph, and the equation in the form $y = mx + b$. Such patterns are ubiquitous in daily life, where constant rates are the norm—coffee for \$8 per pound, 150 calories per glass of milk, a \$2 service charge per call. A person's ability to recognize these linear situations and draw conclusions from them is enhanced as he or she comes to understand the pattern characteristics of linear functions.

In the final lessons of the unit, some nonlinear patterns and functions are introduced as a way to highlight further the features of a linear pattern or function. The linear function with its constant rate of change is contrasted with the nonconstant rate of change of simple quadratic and exponential functions.

Making sense of algebraic notation

> *That which makes "algebraic" the students' thinking is the distinctiveness of the mathematical practice in which they engaged, namely the investigation and expression of the general term of a pattern—something that may not be required at the level of the arithmetic thinking. That is why, for us, the students were already thinking algebraically when they were dealing with the production of a written message... despite the fact that they were not using the standard algebraic symbolism (Radford, 2000, p. 258).*

Some people think that one is not doing algebra unless one is able to write quantitative relationships down in symbolic algebraic notation (e.g., $2x + 3x = 5x$) and demonstrate awareness of the structure and properties of the number system (e.g., commutativity: $ab = ba$). Other people, the authors of this unit among them, disagree with this notion. That is not to say it is unimportant to become more at ease with notation; in fact it is a facility that is essential for numerate adults to develop.

In this unit on algebraic thinking, skill with algebraic expressions and equations develops in a natural way by using students' words, diagrams, graphs, and tables as the basis for translating a situation into mathematical language. It is important, therefore, to connect notation conventions with students' words and thoughts—in other words, the symbols are about something. Students develop a sense of what the symbols mean and how they work as the need arises within the context of each investigation. *Symbol Sense Practice* is provided as a follow-up activity in all lessons so that students can solidify their personal command of the notation.

While some of the power of algebra lies ultimately in the efficiency with which one can manipulate symbols—for example, transform $x(x + 4)$ to $x^2 + 4x$—it is important that such symbolic manipulation be robustly grounded in the meaning behind those symbols. Algebra and its symbols should be demystified. Algebra is not magic; it makes sense, and it can be a useful tool given the proper introduction and development. The hypothesis is that seeking patterns and building rules is a better focus when starting algebra than symbol manipulations because it sets the right tone for expressing generality, a central component of algebraic thinking.

> *When awareness of generality permeates the classroom, algebra will cease to be a watershed for most people; that algebra as it is usually interpreted at school is a dead subject, akin to conjugating Latin verbs and memorizing parts of flowers.... If teachers are unaware of its presence, and are not in the habit of getting students to work at expressing their own generalizations, then mathematical thinking is not taking place (Mason, 1996, p. 65).*

Unit Goals

By the end of this unit, students should be able to

- Identify patterns and predict outcomes in a variety of situations.
- Describe patterns and relationships using diagrams, words, tables, graphs, and/ or equations.
- Understand how different representations are related.
- Recognize the characteristics of linear patterns.
- Use basic algebraic notation.

The Flow of the Unit

The *Teacher Book* provides materials for initial, midpoint, and final assessments (see *Assessments*, p. xxv, for details). During the first three lessons, the emphasis is on tables and rules—that is the organization and use of In-Out tables listing *x*- and *y*-values in order to identify the function or "rule" at work in the pattern or relationship between those values. In *Lessons 4–6*, the repertoire of algebraic representations expands to include graphs. The emphasis shifts to equations in *Lesson 7*. Once established, the foundational skills of using, interpreting, and connecting algebraic representations lay the groundwork for explorations of linearity and its variations in *Lessons 8 and 9*. *Lessons 10–12* then focus on rates of change, both constant rates of change in linear relationships and nonconstant rates of change in nonlinear relationships.

Each lesson has one or more activities or investigations. Allowing time for opening and summary discussions (including time for students to write reflections) and assuming a thoughtful pace, most lessons will exceed an hour. Students' reading and writing levels will have an affect on how long each lesson takes.

In *Lessons 1–3*, students

- Complete and construct tables, write rules that generalize patterns revealed through the tables, and work through a problem using these skills.

In *Lessons 4–6*, students

- Review graphing conventions, graph table data, and compare graphs for positive and negative linear relationships.

In *Lesson 7*, students

- Examine equations by translating rules into equations and connecting equations with graphs, tables, rules, and situations.

In *Lessons 8–9*, students

- Solve problems using a repertoire of representations, as they explore linearity and its constant rate of change.

In *Lessons 10–12*, students

- Contrast linear and nonlinear situations to deepen understanding of constant and nonconstant rates of change as represented in tables, graphs, and equations.

Assessments

Seeking Patterns, Building Rules opens and closes with assessment. In both cases, the *Teacher Book* provides multiple ways to gauge what students know. In addition, the *Teacher Book* includes an interim assessment for students. There are also tools for ongoing assessment in each lesson. These components are described below.

Opening the Unit

Algebra Mind Map: Students record words and ideas they associate with algebra, providing a snapshot of background knowledge. By the end of the unit, their Mind Maps will likely include more **terms**, particularly technical ones.

One of My Own Patterns: Students describe a pattern in their own lives and how it extends over time. They then represent the pattern in a graph, diagram, table, and/or equation.

Initial Assessment: Patterns and Problems: Students self-assess their ability to complete visual and numerical patterns; generate graphs and tables; match tables, graphs, and equations; and complete several symbol sense activities. Then, if they are able to, students complete the tasks. Checked against the *Initial Assessment Checklist*, the *Initial Assessment* will indicate which lessons will be new material and which lessons will be reviews.

Interim Assessment: Using the Tools of Algebra

The *Interim Assessment* covers concepts from *Lessons 1–6* and includes five parts.

Working from a Graph: Students examine a graph to tell the story behind it, and make a table, rule, and equation corresponding to the graph.

Working from a Table: Given a table, students examine the pattern, describe the situation, make a corresponding graph, solve a problem, and write an equation.

Working from an Equation: Students connect a formula to a situation and then generate a table and graph.

Symbol Sense: Students verify mathematical equations and translate words into symbols, and symbols into words.

Closing the Unit: Putting It All Together

Some Personal Patterns: Students pick a personal pattern to represent with a table and a graph, as a rule in words, and as an equation.

Match Situations with Representations: Students match situations with graphs, tables, and equations and then solve three related problems.

Symbol Sense Stations: Students evaluate algebraic expressions and solve one- and two-step equations.

Mind Map and Unit Reflection: Students record words and ideas they associate with algebra. Their Mind Maps will likely include more terms, particularly technical ones, than were in their first Mind Map in *Opening the Unit*. They reflect on what they have learned in the unit.

Ongoing Assessment

Much of the work students do will take place in small groups or pairs. This work must be assessed to determine whether students are absorbing the main ideas and to diagnose difficulties, as well as provide more challenging work to high-level students. The *Looking Closely* section of each lesson focuses teachers' attention on objectives and the corresponding observable behavior. Some teachers read this after they have taught the lesson and return to class prepared to assist students who struggled. Teachers may also revisit ideas that were not clearly understood.

As students gain skills, teachers will want to track their progress and communicate observations to them. Each lesson has a chart called *What to Look For in Lesson #*, with two or three dimensions for assessing students' progress:

Representations—ability to generate and interpret information from diagrams, tables, graphs, equations, and rules.

Connections—ability to link representations through salient features.

Problem Solving—ability to apply representational tools in order to enter and accurately solve problems.

Many people question, "Why study algebra?" As students work through this unit, they will come to understand how useful algebra can be. They will share knowledge gained in and out of the classroom to become proficient at using algebraic tools and concepts.

Facilitating Opening the Unit: Seeking Patterns, Building Rules

What makes something a pattern?

Synopsis

This session assesses students' familiarity with algebra and algebraic representations—tables, graphs, and equations. The activities establish baseline information about students' background knowledge and set the stage for later work with patterns.

1. Together, students begin to think about and discuss algebra, as they share past experiences, complete an algebra Mind Map, and start the class algebra vocabulary list.
2. The class considers a visual pattern, then brainstorms a list of other possible patterns before composing a preliminary definition of "pattern."
3. Individuals develop words and representations to describe a personal pattern involving numbers.
4. Students complete a self-assessment based on several problems. They complete all or some of the problems in the *Initial Assessment*.
5. A discussion of the *Unit Goals* and personal goals reinforces the scope of the unit. Tips for success are offered.

Objectives

- Share past experiences with algebra and existing notions of algebra
- Begin a record of vocabulary useful to the study of algebra
- Look for patterns and relationships in a variety of representations
- Identify and represent a personal pattern in words and diagrams, graphs or equations
- Understand the scope of the unit

Materials/Prep

- Markers
- Easel paper
- Rulers

Make a copy of *What to Look For in Opening the Unit*, p. 9, to record your observations.

Prepare a class vocabulary sheet to be posted at the front of the class and added to throughout the unit.

Prepare a drawing or other visual of the Mind Map (save results for later comparisons).

Prepare presentation slides of the *Initial Assessment*, pp. 175–178.

Make a copy of the *Initial Assessment*, pp. 175–178, for each student.

Make a copy of the *Initial Assessment Class Tally*, p. 179, to use in analyzing the *Initial Assessment.*

Make copies of the *Initial Assessment Checklist*, pp. 180–181, one per student, to use for individual assessments.

Prepare a presentation slide of "Stick Figures," *Blackline Master 1*, or draw something similar on the board for the whole class to see.

Opening Discussion

During the next few weeks, students will develop the skills and concepts that enable them to solve problems using various algebraic representations. They will come to better understand graphs, tables, and equations, and the patterns they communicate.

Many students find the prospect of studying algebra intimidating. Begin by inquiring about students' past history with algebra so you can get an idea of their background knowledge. Tell your students:

We are about to embark on a journey to learn some algebra, but before we do, I would like to hear what you already know about this subject.

Find out who has studied algebra, where, and when. If few or none of your students have studied this subject, tell them you would still like to know their ideas about it.

By the end of *Opening the Unit*, you will have several pieces of evidence of students' prior knowledge of algebra:

- *Algebra Mind Map (Activity 1)*
- *One of My Own Patterns (Activity 2)*
- *Initial Assessment (Activity 3)*
- The start of a class vocabulary list

Activity 1: Algebra Mind Map

Start the Mind Map by writing "Algebra" on easel paper or projecting the word on a screen. Ask:

What comes to mind when you think of the word "algebra"?

Take a few ideas. Model clustering them, linking related ideas with lines, and then ask students to record their own ideas on p. 2 of their books. After a few minutes, ask students to write one or two of their ideas on sticky notes (or, if a slide is used, to call them out for you to record). Invite students to add their ideas to the class Mind Map, creating new clusters as needed.

As terms related to algebra arise, start a class vocabulary list for algebra. Do not worry about formal, dictionary-style definitions; use what the students themselves say. You will want to add and refer to this class vocabulary list, so write it on easel paper and post it at the front of the class for the duration of the unit. Students can take notes in the vocabulary section of their books, starting on p. 155.

Save the individual Mind Maps as well to compare them to responses at the completion of the unit.

Activity 2: One of My Own Patterns

In this activity, students represent a personal **pattern** first with words and then by creating a table and a numerical equation. Take some time to generate thinking about patterns. As an introduction to the practice of using a pattern for prediction, offer students a chance to work with the "Stick Figures" pattern.

Project an image of *Blackline Master 1* or draw on the board the following:

Ask:

What comes next here? How do you know?

Allow time for students to share their opinions, make their cases, and come to the agreement that the next figure would have its hands *down*. Then ask:

What would the rest of the pattern look like?

Pay attention to students' reasoning. Some students may say, "Even up, odd down." Others may find it difficult to see and describe the pattern. Ultimately, there should be agreement that odd-numbered groups of figures have their hands down, and even-numbered groups have their hands up: 1 down, 2 up, 3 down, 4 up, 5 down, and so on. Verify by asking:

What would the 13th group look like? What about the 536th group?

Make a statement that focuses on prediction to close:

The wonder of mathematical patterns means that once we define the rule at work in the pattern, we can predict what will happen far ahead.

Follow up by saying:

Patterns are everywhere around us. This drawing shows one kind of pattern. What other patterns do you notice in everyday life?

Record students' answers on easel paper or on a slide.

Categorize a few of the patterns. If students list only visual patterns, ask for examples from nature, social studies, mathematics, and from personal routines. Extend the discussion, and work toward a definition of pattern by asking:

If these are examples of patterns, what makes something a pattern?

Ask what the patterns have in common. Record students' responses on the board. When terms like "repeating" or "over and over again" appear, ask whether someone can summarize to create a definition of "pattern." Once the class agrees, write the definition on the class vocabulary list.

Close the discussion by asking:

How can knowing or recognizing a pattern help you?

Here you are looking for the idea that patterns help us to *predict* what will happen in the future or what will come next, or to replicate something.

Starting the Main Activity

Introduce *Activity 2* by saying:

We will take a few minutes to explore one of your own patterns. What are some of your daily patterns that might involve numbers?

As a class, brainstorm a list of ideas (see *Opening the Unit in Action*, p. 13, for one class's list). Guide further thinking with questions such as these:

Which of these patterns can you picture in a graph? How about in a diagram or picture? In a chart of information?

For which of these patterns would you find a prediction interesting?

Let students share their thinking, but do not belabor this analysis. Refer to *One of My Own Patterns* (*Student Book*, p. 3), and review directions. Suggest students choose a new personal pattern or one from the class list. Encourage students to use complete sentences when describing their pattern in words and then to create an alternative representation.

Once everyone is done, have students share their representations. There are several ways you might choose to orchestrate the sharing. Some ways that teachers have done this successfully are

- Every student makes a poster on easel paper of his or her pattern in words with the nonverbal representations. Then students make predictions about each other's patterns over a certain period of time, e.g., a year.
- Students pair up and exchange their nonverbal representations. Each tries to predict something from his or her partner's representation.
- Students who finish early copy their work on easel paper to share with the whole class.

Whichever way you facilitate the sharing, ask:

How does the representation help you predict?

What would come next in this pattern? (Choose one student's pattern.)

To summarize, point out the different types of representations used (e.g., pictures, graphs, equations). Explain that with some of the representations it is easy to predict outcomes. Add that the ability to predict what comes next in a pattern or what a future result will be is an algebra task with which they will become very familiar.

Tell students they will return to some of these real-life situations later, seeing with new eyes patterns, tables, graphs, rules, and equations that relate to them.

Activity 3: Initial Assessment—Patterns Problems

Because the *Initial Assessment* can look intimidating, it is recommended that you begin by explaining that this is a test where students first check off whether they feel they can do, don't know how to do, or are not sure whether they can do the problem. Tell your students that the results will help you decide what is most important to focus on in the unit on data and graphs.

The assessment can be run as a group activity by projecting the problems on a screen and having the class read the problems together. Distribute a copy of the *Initial Assessment* to every student so that individuals can assess their confidence in their ability to solve the problems. Students can read silently or you can read to them. Keep the momentum going, reading the problem, and then moving on to the next one.

The *Initial Assessment* is a flexible tool. You can gain good information about students' skills simply by asking them to self assess, indicating their confidence levels on the problems by marking them "can do," "don't know how," or "not sure" *without solving* them. Alternatively, students may solve the problems, starting with those that they feel they can solve and attempting those they are not certain about. Establish a firm time limit for this. Then collect the assessments and analyze them later. Students' work on the problems will provide a point of comparison for assessing their progress when they have reached the *Midpoint* and *Final Assessments.*

See the *Initial Assessment Checklist,* p. 180, for information on instructional decisions.

Summary Discussion

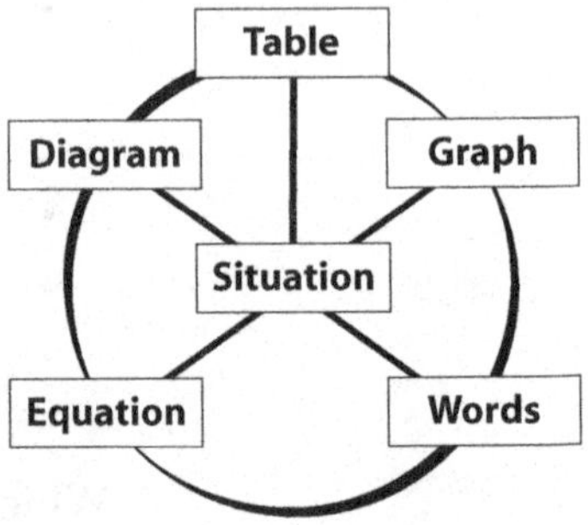

Introduce the circle diagram on p. 5 of the *Student Book* to highlight the interconnections of the representations and the central situation.

Sum up the introduction to the unit:

One purpose of this unit is for everyone to appreciate the possibilities of algebra. That's why the unit starts with patterns in our own lives that can be described in words and in algebraic ways, for example with graphs and tables.

Another purpose is to take the mystery out of algebraic letters and numbers so they make sense. Using algebra to make sense of the world is the underlying goal of the unit.

Discuss some everyday situations where algebra might have relevance for students—college, the GED test, high school diploma, work (almost every business uses spreadsheets such as Microsoft® Excel where algebraic formulas lie hidden behind the cells). Review the *Unit Goals*, p. 4, and ask:

What do you understand the *Unit Goals* to be?

Can you think of anything else you need or want to know?

If there is time, orient students to their student books.

Have students answer the *Reflections* questions, p. 158, and make entries into their *Vocabulary* list, p. 155.

Looking Closely

Observe whether students are able to

Share past experiences with and existing notions of algebra

The Mind Map will provide some insight into students' understanding of and attitudes toward algebra. Some students' understanding will be more developed than others'; these students might say "solving equations," "using formulas," or "working with positive and negative numbers," rather than "Algebra has lots of letters." Some might voice affective issues, such as "It's complicated," or "I don't get it." These varying responses will guide your approach with students.

Give support and encouragement where needed, and challenge those who are prepared.

Begin a record of vocabulary useful to the study of algebra

Introduce the *Student Book* format, so students can find the *Vocabulary* list at the end of their books. Encourage them to make notes of new terms, using their own words. When students ask for the meaning of words such as "sequence" or "prediction," provide them. However, teaching vocabulary is not the intent at this point. You want to see whether terms that surface are familiar to students, whether they are able to use them accurately and define them for others.

When the class defines a word together, such as pattern, their understanding of the word, rather than a dictionary definition, should guide the explanation.

Look for patterns and relationships in a variety of representations

The *Initial Assessment* offers many opportunities for students to search for patterns and relationships. Are students able to predict the next shape in a visual pattern? Can they explain why it comes next? Can they predict the next number in a number sequence? Guide students to pay attention to visual features. Ask them what is changing or what they can count to determine whether an increase or decrease is happening. With number patterns, ask them what the increments for the changes are. Sometimes students do not know where to begin; you can help by asking guiding questions without providing the answers.

With the "Stick Figures" pattern, those who only see up and down (attaching no numbers or number types to the patterns) need to see graphic representations of figures to determine future events. They might benefit from questions like, "If we label the figures with numbers, do you see a pattern now?" Those who see that the pattern involves odd and even numbers are able to generalize and make predictions.

Identify and represent a personal pattern in words and diagrams, graphs, or equations

The choice for a personal pattern often influences ability to represent that pattern mathematically. Students sometimes choose a pattern of daily activity that precludes graphing and equation writing. Other times, they see a pattern as repeating but not accumulating. Guiding students to focus on predicting future events can help them see the difference between representations that lend themselves to predictions over time and those that do not.

Like the Mind Maps, there are no right and wrong answers with personal patterns, only assessment information for you. Can students describe a pattern accurately with words? Can they make a graph? What are the features of the graphs they make? Do they depict their patterns with graphic diagrams or symbolic ones? Are they familiar enough with mathematical patterns to choose an accumulating pattern for the activity? Do they simplify their pattern into an algebraic formula or equation? All of this can be assessed by carefully studying the personal patterns students choose and how they are presented.

Understand the scope of the unit

The assessment session concludes with a presentation of the *Unit Goals* and asks students to state their own goals. Use the students' goals to inform your instructional goals. The final page offers some tips for success and presents a diagram that communicates the unit's organizing principle: The pattern is key, and a person can model patterns with several tools: diagrams, tables, graphs, verbal rules, and algebraic equations.

WHAT TO LOOK FOR IN OPENING THE UNIT	WHO STANDS OUT? (LIST STUDENTS' INITIALS)			NOTES FOR NEXT STEPS
	STRONG	ADEQUATE	NEEDS WORK	
Background knowledge • Uses algebraic vocabulary • Reports experience with algebra • Reports comfort level with assessment tasks • Constructs a graph from a table or written pattern				
Identifies and represents a personal pattern • Clearly, accurately describes a personal pattern ® In words ® In other forms (graphs, tables, equations)				
Extends a visual pattern • Identifies the "Stick Figures" pattern • Accurately predicts the 13th or 536th cases				
Looks for patterns and relationships • Identifies changing elements in visual patterns				

Rationale

Before teaching a unit, it helps to gain a sense of what background knowledge and skills students already possess. You want to understand what students know about describing patterns, writing equations, and constructing representations. By exploring personal and visual/geometric situations that encourage students to search for and identify patterns, you set the stage for more formal work with algebraic representations. You also lay the groundwork for further exploration of patterns that represent linear relationships. The *Initial Assessment* can provide you with a sense of both the class's and individual's skills and abilities. As the class works through the activities, these will become more apparent.

Math Background

Any situation may exhibit several patterns occurring all at once. Take, for example: "I drink eight glasses of water a day." One person may focus on the repetitive pattern of *the same thing happening day after day after day* and is likely to show representations such as

8, 8, 8, 8, 8...

Mon	8
Tues	8
Wed	8
Thu	8
Fri	8

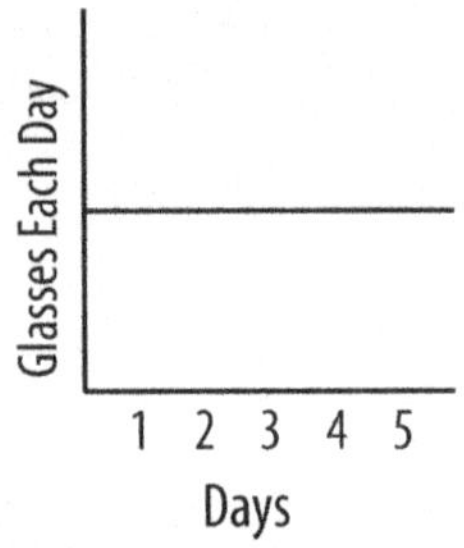

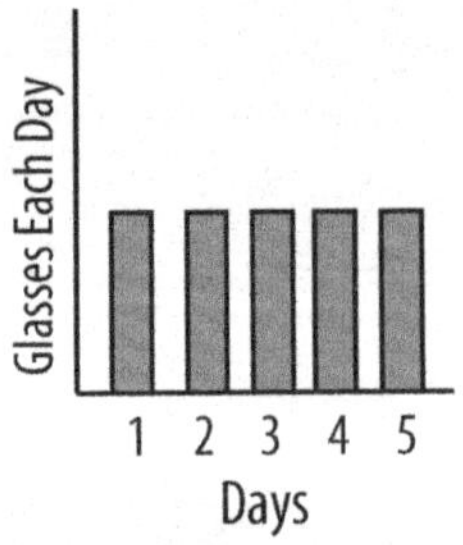

This focus enables us to predict the number of glasses of water consumed on any particular day (where eight is the number of glasses each day).

Another person may focus on *how the total number of glasses accumulates over time.* Representations that focus on how the total number grows over time when you drink eight glasses of water a day might look like this:

8, 16, 24, 32...

Mon	8
Tues	16
Wed	24
Thu	32

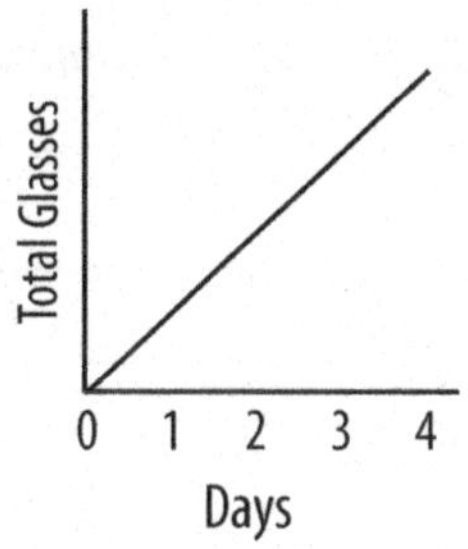

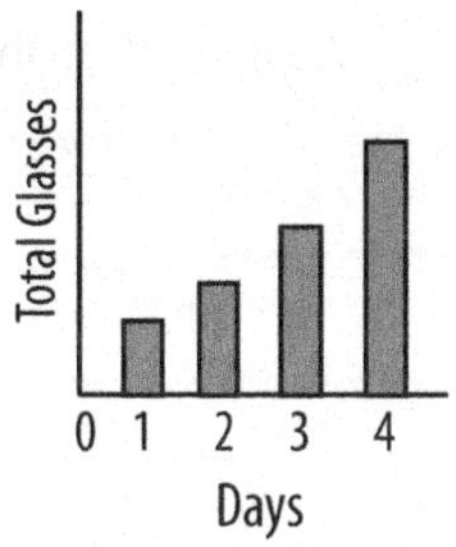

Both of these patterns provide information about a situation, but the second (looking at accumulation) enables us to predict the total number of glasses of water consumed after any number of days ($n = 8d$, where n is the number consumed by the dth day). This unit focuses on growing patterns and on how amounts accumulate, e.g., how many calories you burn after so many minutes of exercise. Nearly all the patterns in the unit are linear.

Facilitation

Remember that this lesson is an initial assessment. The intent is *not* for you to teach students how to graph or make a table to solve a problem; write a formula to describe a pattern in their lives; or learn how representations are interrelated. The intent is to discover which representations the *students* are familiar with, and to what extent. Listen carefully to what students say, and closely observe what they write.

The goal of the entire unit is for students to learn to generalize and connect patterns in algebraic rules, but for now, you just want to get a sense of where the class and individuals are in their development of algebraic thinking. Teaching students to solve the problems at this point may undermine the sequential development of skills and concepts.

The *Initial Assessment* offers a variety of tasks for students to use for self-assessment; later they may attempt to solve the problems. The self-assessment gives you an idea of students' pre-existing knowledge and their confidence levels. Parts of the assessment will be challenging for even the most advanced students, especially if time constraints are set. Probe for the reasoning students used to make specific choices, but do *not* show anyone how to do the problems. Instead, encourage students to keep an eye on the numbers and how they are related. Remind them that you are trying to find out how much they already know about algebra so you can plan what to teach.

Withhold judgment when the class is sharing from *Activity 2: One of My Own Patterns*. Again, you are seeing what people know and how they use their knowledge, not looking for a preconceived answer. At the close of the unit, you will return to some of these personal patterns and see how students' learning affects their responses to similar questions.

Making the Lesson Easier

Do not have students solve all the problems in *Initial Assessment.*

Making the Lesson Harder

Assign all problems in *Initial Assessment.*

From the "Patterns" Discussion

The teacher for a class that listed a great variety of patterns asked the students to differentiate those that involved numbers from those that did not. This helped her begin a discussion on patterns and their purpose.

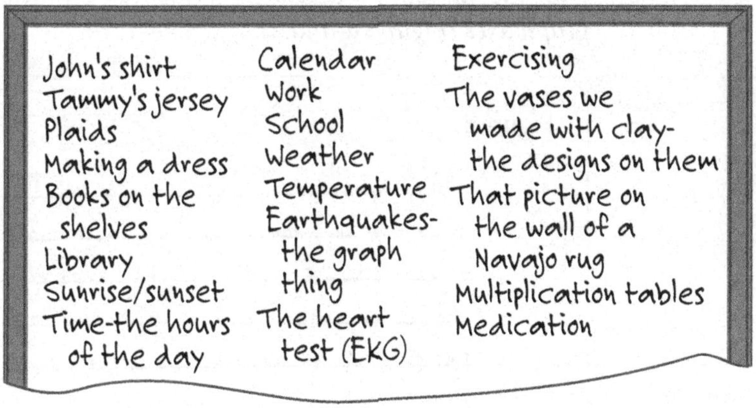

Students at the Pioneer Valley Adult Education Center
Northampton, MA

From "Stick Figures"

Pushing for predictions helps students analyze relationships.

In "Stick Figures," all but Cathy recognized the pattern as "1 down, 2 up, 3 down, 4 up," and all knew the next group would be "5 down." Cathy did see the pattern when Jose and Anna described it, with the others joining them. They left it at that until I asked, "If I said there were 13 stick figures, would you know whether their arms were up or down?"

We decided that, in Juan's words, "Odds were down and evens up." This meant we could predict whether the 536th group would have its arms up or down—definitely up.

Tricia Donovan
Pioneer Valley Adult Education Center, Northampton, MA

From *Activity 2: One of My Own Patterns*

In pilot classes, we observed people having little or no problem expressing a personal pattern in words, but there were a lot of blank looks when we asked them to represent that same pattern with a diagram, graph, table, or equation. Once they did get started, the set of representations were communicative, yet often unconstrained by conventions. The following illustrate the range of alternative representations offered in the assessment activity.

Note the inattention to scale, as evidenced by unequally spaced intervals on the graph axes (Figures 1 and 2).

Figure 1

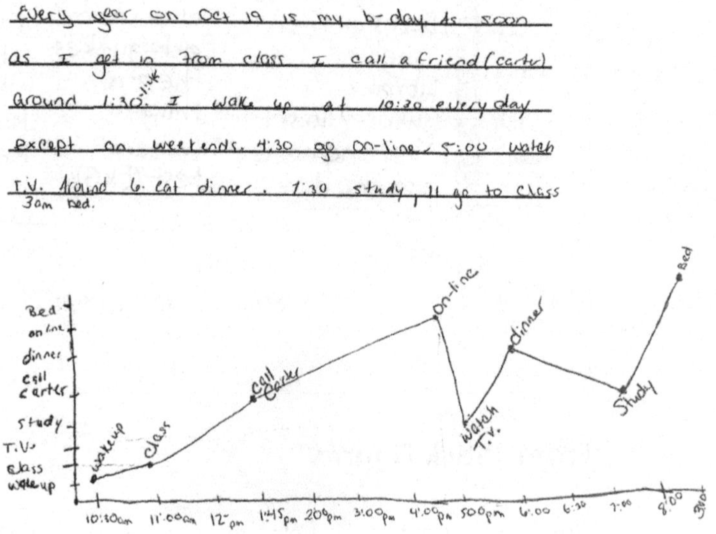

Figure 2

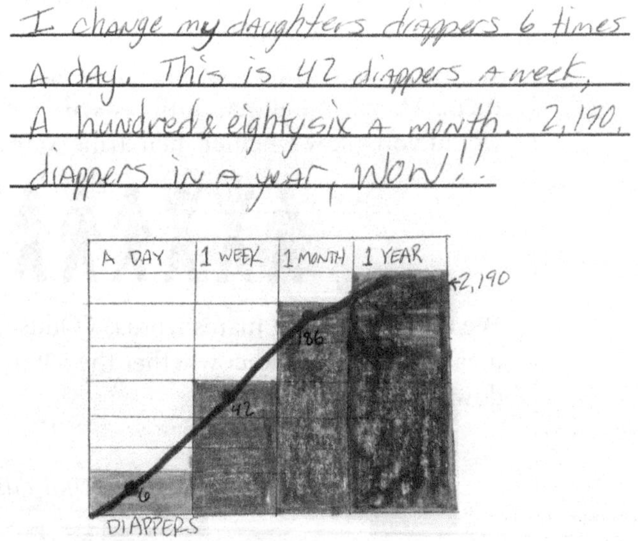

This student's table (Figure 3) corresponded well with her verbal pattern; however, a two-column table would have sufficed.

Figure 3

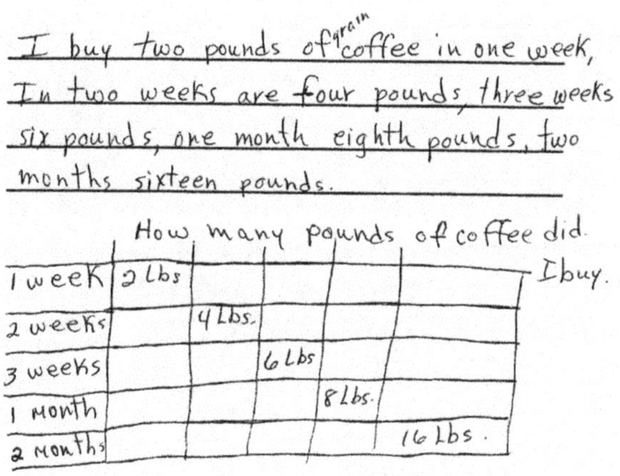
I buy two pounds of grain coffee in one week, In two weeks are four pounds, three weeks six pounds, one month eighth pounds, two months sixteen pounds.

How many pounds of coffee did I buy.

1 week	2 Lbs				
2 weeks		4 Lbs.			
3 weeks			6 Lbs		
1 Month				8 Lbs.	
2 Months					16 Lbs.

The students at a more basic level seemed unequipped to produce a graph or table and, like the student below, drew a literal picture of a single event (Figure 4).

Figure 4

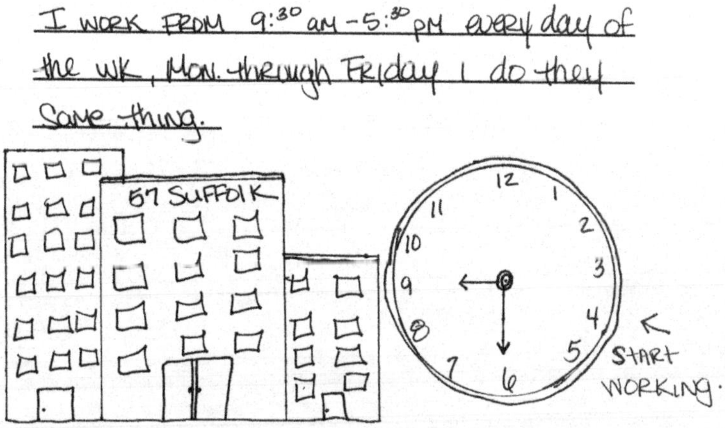

Compilation of student work from pilot classes

FACILITATING LESSON **1**

Guess My Rule

How can I explain the rule behind this table of numbers?

Synopsis

This lesson uses a game, *Guess My Rule*, to introduce In-Out tables as tools for stating a rule that describes the relationship between two variables.

1. The class uses a table (*The Waitress's Problem, Blackline Master 2*) to problem-solve and discusses recognizing a pattern and stating a rule for it.
2. The class plays *Guess My Rule*, with you presenting tables from which students generalize rules.
3. Pairs of students create tables and write rules and equations for further rounds of *Guess My Rule*.
4. Student pairs share their verbal rules and equations, and the class tries to match them up. They then match rules to tables.
5. The class summarizes what they learned about seeking patterns in tables and writing rules in words and symbols.

Objectives

- Use In-Out tables as tools that show patterns and relationships between two variables
- Identify numerical patterns that involve addition, subtraction, multiplication, division, or a combination of operations, and represent them in words
- Represent patterns in equations that use algebraic notation

Materials/Prep

- Colored markers
- Scissors
- Tape

Bring in examples of tables you find in newspapers, bills, etc. Good examples are wallet-size sales tax tables, federal tax tables, and mortgage tables.

Make a presentation slide or a easel paper copy of *The Waitress's Problem* (*Blackline Master 2*) and post for the whole class to see; or make photocopies and distribute one to each student.

Make slides or easel paper copies of the tables from *Activity 1: Guess My Rule* (*Blackline Master 3*).

Prepare wide strips of easel paper for groups to use in *Activity 2: Making Tables, Rules, and Equations*, three strips per group.

A note on vocabulary: Throughout the unit, introduce vocabulary by connecting it to students' everyday experiences. The class can develop some working definitions that they will refine as their understanding develops. The vocabulary words to focus on in *Lesson 1* are the following:

- A **table** is a chart with information arranged in rows and columns.
- An **In-Out table** has a rule built into it.
- A **variable** is an amount whose value varies or changes. The In-Out tables we work with here have two variables.
- An **equation** is composed of two mathematical expressions that show how two quantities are equal. An equation always uses an equal sign (=).

Opening Discussion

Connect students' everyday experiences with numerical tables. Project or otherwise share some examples of tables, such as sales tax tables, mortgage tables, or population tables; or make photocopies of the tables and distribute a set to each student. Elicit other examples of numerical tables from the class. Ask:

Where else have you seen tables like these?

Display or distribute copies of *The Waitress's Problem* (*Blackline Master 2*) as an example of a situation where a table is used informally. Ask a volunteer to read the problem aloud. Follow up by asking:

Who can continue the table, so we can see the answer to Kim's question?

Ask a volunteer to extend the table, and seek agreement on accuracy from the class. Then ask other volunteers:

How do you know which number comes next?

Blackline Master 2

The Waitress's Problem

Kim was waitressing for the first time. She did not know what to expect for tips. A veteran waitress told her, "Usually folks here give a 25% tip." Kim must have looked confused; the experienced waitress ripped off a piece of paper from her check pad and scribbled on the back:

Bill	Tip
$1	$0.25
$2	$0.50
$3	$0.75

"It just goes like that," she said, as she handed the piece of paper to Kim. Kim thanked her, thinking, "Yeah, but what should I expect to get for a bill that totals $7.00?"

Are there any other ways?

Listen for explanations of prediction strategies and comment on the many ways used to describe the pattern. Summarize:

What would you say is the rule behind the waitress's table?

Write down the rules students give as answers. They may start with partial statements that lack precision, but push them for fuller statements and more precise language. For example, if a student says, "The rule is 25 cents," write down "25 cents." Then ask, "Do you mean 25 cents is the tip for a bill of any amount?" A more complete response might be, " Multiply the number of dollars in the bill by 25 cents." Push for a statement that relates the two variables (bill amount and tip) by asking: "So what would that give you?" A complete statement of the rule in words could be: "Multiplying the number of dollars in the bill by 25 cents will give you the right amount for the tip."

Explain that this class session is about looking for rules in tables and practicing writing those rules in simple terms and in equations with algebraic symbols.

Activity 1: Guess My Rule

The objectives for this game are to

- Explore strategies for determining patterns by observing the values in the columns and rows of a table of values.
- Streamline and clarify verbal rules.
- Share ways of writing equations that represent rules.

As a class, review the directions in *Guess My Rule*, (*Student Book*, p. 8). You may want to post the list of actions students need to take: look for patterns; suggest a new pair of numbers; write the rule in words; and write the rule as an equation.

Have handy your copies of *Guess My Rule Tables* (*Blackline Master 3*). Begin the game by posting one table from *Blackline Master 3*. Be sure everyone understands what is expected. Check to see that all students have responded correctly to this first table before moving on to the other tables.

Table 1

x	y
2	6
3	9
4	12

$y = 3x$

Table 2

x	y
3	120
5	200
7	280

$y = 40x$

Table 3

x	y
56	7
96	12
104	13

$y = \frac{x}{8}$

Table 4

x	y
1	
18	24
21	27
	40

$y = x + 6$

Table 5

x	y
2	0
3	
4	2
	12

$y = x - 2$

Table 6

x	y
1	3
2	5
4	
	17

$y = 2x + 1$

Table 7

x	y
1	4
2	6
4	10
	220

$y = 2(x + 1)$ or $2x + 2$

Pose questions such as

Who thinks they have a pair of numbers that fits the pattern?

Record some pairs of numbers and ask the class:

Do you agree? Why?

Repeat this procedure two or three times. Then ask:

What do you think the rule is? What did I do to the *x*-value to get the *y*-value?

Is there another rule? How can we check?

You want students to analyze their pattern-seeking processes, so pose additional questions such as

What clues helped you decide on the pattern? Tell us step by step how you found the pattern.

Ask students to write rules in their own words to represent the relationships they see. Help them until each statement clearly expresses the mathematical steps taken to change *x*-values to *y*-values. Next ask the class:

How would you write an equation that shows how you find *y* when you know what *x* is?

Students may fumble at first, as they convert their verbal rules to symbolic ones. Focus on the term **equation**, asking for examples ($4 + 3$ is not an equation; $3 + 4 = 7$ and $7 = 3 + 4$ are). Notational issues that might come up are

Does $4x = y$ mean the same as $y = 4x$?

Does $4 \times x = y$ mean the same as $4x = y$?

As you introduce patterns that involve addition, subtraction, multiplication, division, or a combination of operations, discuss students' strategies for determining those new patterns. Remind students that not all patterns involve just one mathematical operation.

The class establishes rules in words and in algebraic equations for the first seven tables before you summarize the pattern types they looked at: multiplication, addition, subtraction, two-step, and situations where *y* equaled a number minus *x*.

Activity 2: Making Tables, Rules, and Equations

Say:

Now it's your turn to make a table and a rule for it.

Have students pair up. Refer them to *Activity 2: Making Tables, Rules, and Equations* (*Student Book*, p. 9). Review directions.

Note where students begin—with a verbal rule or an equation? Tell those students trying to start with making a table that it helps to have a rule in mind first. You might also say:

If I put this number in for *x*, I should get this number for *y*. Do you agree?

Allow about five minutes for students to complete their tables and verbal and symbolic rules. Encourage them to create challenging rules.

Distribute three easel paper strips to each pair of students. Have pairs copy their table, rule in words, and equation, each on a separate strip of paper. Ask for three volunteers. One should collect all the strips with the tables, one all the strips with the verbal rules, and one all the strips with the equations.

Create three sections on the board or wall, labeled "Problem in Words," "Equation," and "Table." The volunteers post each type of representation under the correct heading but in random order. The board should look something like this:

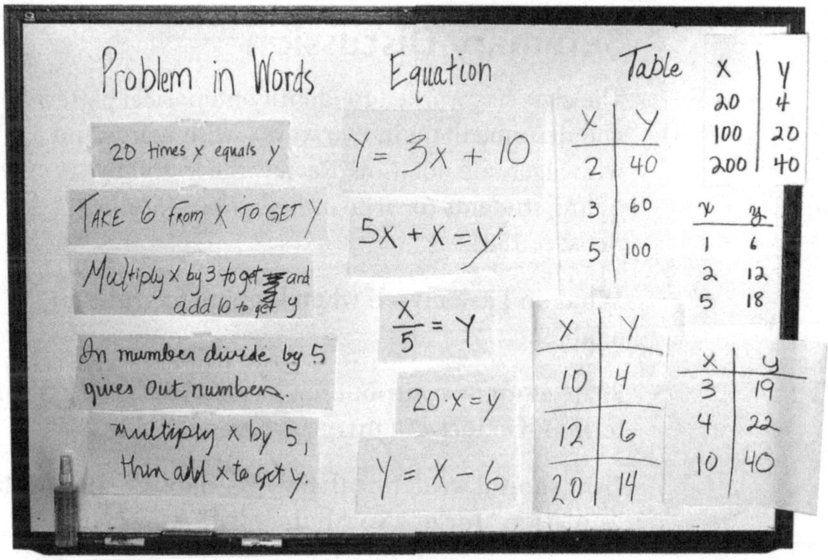

Ask the class to match the tables, rules, and equations.

Draw attention to the section with equations:

Where is a verbal rule that matches this one?

Match symbolic and verbal rules one at a time, moving them on the board to be adjacent. Discuss what makes matching easy or hard. Solicit alternative ways to write rules in words or in symbols to clear up misconceptions about notation and verbalization.

Heads Up!

This discussion creates many opportunities to discuss notation conventions. Comment on these when useful, but focus on students' learning to recognize and describe patterns.

Next, match each rule set with its correct table, moving them to be adjacent on the board. You can develop metacognition by asking for a review of processes:

How do you know the rules and tables match? What do you look for? What do you ask yourself?

Students should support statements by connecting them to the table data.

Introduce the term "variable." Say:

***x* and *y* are called variables because their values can change or vary.**

To set the stage for later work, tell students:

When you write a rule using variables, such as *x* and *y*, you are generalizing about the pattern, making a general rule to follow.

Generalizing in these ways (point to verbal and algebraic rules) allowed you to *predict* future values.

Summary Discussion

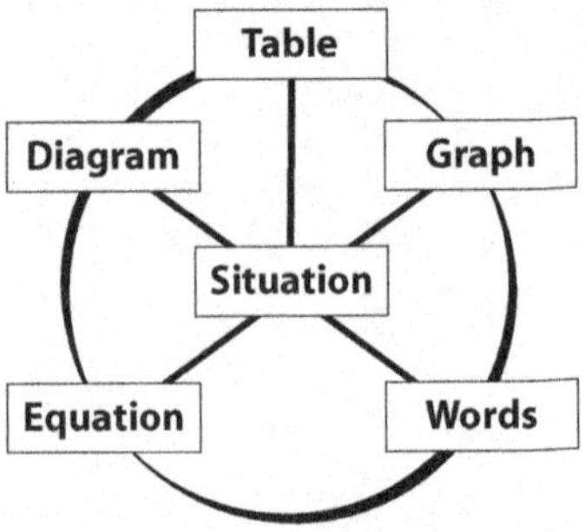

The class has worked to identify numerical patterns and describe them in two ways—with words and with algebraic notation. Review the lesson by asking students to write in *Reflections* (*Student Book*, p. 158):

What do I ask myself when looking for rules in a table?

Share as many contributions as possible. Then ask for working definitions of terms introduced in this class session. Ask:

How should we define the words "table," "In-Out table," "variable," and "equation" for our vocabulary list?

Add the terms and their definitions to the class vocabulary list, and ask students to copy these into their *Vocabulary* (*Student Book*, p. 155).

Practice

Guess More Rules, p. 10
Gives students more opportunities to write the rules behind tables in words and as algebraic equations. Assign to everyone. The practice is useful, and you will be able to assess individuals' skills.

Fill In the Values, p. 13
Asks students to complete tables.

The Rule Story, p. 15
A cloze exercise with contextual practice of terms and rule generation.

Symbol Sense Practice

How Many Ways?, p. 16
Provides practice with multiplication notation.

Four Ways to Write Division, p. 17
Provides practice with division notation.

Extension

Building Tables from Rules, p. 18
Offers work with more complex numbers where students start with the equation and create tables.

Test Practice

Test Practice, p. 19

Looking Closely

Observe whether students are able to

Use In-Out tables as tools that show patterns and relationships between two variables

Students should come to understand that something happens to the numbers in the In column (x) to produce the numbers in the Out column (y). If students have trouble with these tables, offer examples of tables that list terms sequentially (though they do not always have to start 1, 2, 3). This practice is especially useful when introducing new types of patterns. Conversely, create randomly listed values to address an over-reliance on seeking patterns by only looking vertically at values.

You may also want to introduce a three-column table with space in the middle column for the rule, so students can understand that something happens *between* the In and Out numbers.

IN (x)	RULE: ADD 7 TO EQUAL	OUT (y)
4	+ 7————————> =	11
7	+ 7————————> =	14

The terms "column" and "row" can be confusing; clarify them early in the lesson.

Identify numerical patterns that involve addition, subtraction, multiplication, division, or a combination of operations, and represent them in words

Remind students of their definition for "pattern"—things that happen over and over again, or repeat. To identify patterns, they must look for something that happens repeatedly. Admittedly, this goes slower when students are not fluent with addition or multiplication facts; if patterns are not listed sequentially in a table; or if patterns involve more than one step. The act of seeking the pattern, however, motivates students to think about representing it in numerical terms.

When patterns are not evident to students, several steps can be taken. Leave two-step patterns for later. With one-step patterns, ask whether the Out number increases (indicating multiplication or addition by whole numbers) or decreases (indicating subtraction or division).

When introducing two-step patterns, offer one pair of numbers (e.g., 5 and 14) and ask for several possible ways to get from the first value to the second (add 9; multiply by 3, then subtract 1; multiply by 2, then add 4; etc.). Then ask students to eliminate rules that do not apply to a second pair of numbers, e.g., 7 and 20.

Identifying patterns is the key to stating rules. Students need to state rules in their own words to make sense of the pattern. It is equally important that people share various ways to accurately state the rules.

Represent patterns in equations that use algebraic notation

Translating the rule in words to algebraic symbols takes practice. Writing rules for patterns in symbolic notation is a skill that will develop over time.

If there is confusion between In and Out numbers, or between x and y, work on clarifying verbal rules so that changing the words to symbols is less difficult. If students write "$3y = x$" for a table representing the pattern $3x = y$, compare the two equations by restating both in simple terms. Ask:

In the table, do you multiply the In number by 6 to get the Out number or do you multiply the Out number by 6 to get the In number?

Facility with notation conventions develops over time. At first, students may write "4 times x" as "4 x x." In this early lesson, the meaning is what is most important, so using the multiplication sign is acceptable. However, introduce the algebraic convention that the multiplication sign is not usually used, but it is understood that $4x$ means 4 multiplied by x.

WHAT TO LOOK FOR IN *LESSON 1*	WHO STANDS OUT? (LIST STUDENTS' INITIALS)			NOTES FOR NEXT STEPS
	STRONG	ADEQUATE	NEEDS WORK	
Representations Tables • Recognizes patterns • Identifies relationship between *x*- and *y*-values • Predicts missing values correctly Rules • Accurately describes relationship between *x*- and *y*-values in words • Accurately writes relationship between *x*- and *y*-values as algebraic equations • Uses algebraic notation conventions for all four operations				
Connections • Identifies table data patterns related to aspects of rules (where to place the "3" in a table representing $y = 3x$, for instance) • Connects features of the equation to the verbal rule				

Rationale

Students who are able to detect a pattern in a two-column table of "naked numbers" and describe the relationship between the two variables possess valuable problem-solving skills. By providing the opportunity to practice reading a table, this lesson facilitates understanding of the *x*-*y* relationship fundamental to understanding functional relationships in algebra. Learning to describe relationships precisely is essential, and the use of In-Out tables allows students to focus solely on the *x*-*y* relationship without being distracted by extraneous information.

Math Background

Function tables, or In-Out tables, are a common tool used by mathematicians and others to understand the nature of a pattern by looking at numerical examples. All the patterns here are functions, in that for every x there is one and only one y.

Facilitation

Students need time to examine the tables to be able to see the patterns. Allow plenty of time to think. It is important to present the *Guess My Rule* tables in an order that reflects growing complexity and allows students to encounter different types of linear patterns. Bearing this in mind, allow more time for students to work with the later, less transparent patterns.

Making the Lesson Easier

Focus on writing verbal rules, not equations. Supply copies of the 1–12 multiplication tables. Connect the variables to real-life contexts if students are finding the activity too abstract.

Making the Lesson Harder

Discuss the relationships between the notation in the various equations. For example, how does $y = 2x + 1$ relate to $x + x + 1 = y$? Or, how does $x + x + x + x + x + x = y$ relate to $y = 6x$? How does $2(x + 1)$ relate to $2x + 2$?

Challenge students to come up with different tables; then try to think of a situation that might apply to one of them.

Make the numbers harder. Introduce patterns with fractions and decimals.

Make the relationships harder. Share tables with two-step patterns involving division, exponential or square root relationships, or negative numbers (from which students might induce rules about integers).

Students express verbal and symbolic rules in different ways. Some samples of what students wrote in this first class follow.

For a table representing $y = 4x$, students came up with

"Times *X* by four"

"$X \bullet 4 = Y$"

"I find the problem by multiplying by 4 to each number from the *y* table."

"I started by knowing 4 x 4 = 16 so I multiply $X \bullet 4 = Y$."

For a table representing: $y = 2x + 1$, responses included

"Times by 2 and add one."

"$X \bullet 2 + 1 = Y$"

"Double *X* and add one to get *y*."

"$2 \times X + 1 = Y$"

"You have to add *X* by the same number plus 1. $X + X + 1$"

"When you have a number in the *x* column, times it by two and add one, and you will have the answer."

"$(x) \times (2) + (1) = y$"

Students at the Harvard Bridge to Learning and Literacy Program, Cambridge, MA; HALO Learning Center, Holyoke, MA; Mt. Wachusetts Adult Education Center, Gardner, MA; and Career Link, Lancaster, PA

FACILITATING LESSON 2

Banquet Tables

How many tables will I need for a crowd of any specific size?

Synopsis

The banquet tables problem provides students with a situation that leads to a rule for problem solving.

1. Pairs work on the banquet tables problem.
2. Volunteers share their resulting diagrams and tables. Students explain how they figured out the 400th arrangement of tables, which leads to a discussion about the rules embedded in the problem.
3. The class considers the usefulness of diagrams, tables, and verbal statements in building rules. They begin another problem similar to *Banquet Tables.*

Objectives

- Use objects, diagrams, tables, and rules to understand and describe a pattern in a situation
- Keep track of problem-related data in a table

Materials/Prep

- Colored markers
- Easel paper
- Square tiles or sticky notes

Opening Discussion

You might warm up the class by doing two or three quick rounds of *Guess My Rule*. Then mention *The Waitress's Problem* and how the table was extended to figure out the tip Kim could expect to get on a $7 restaurant bill:

Tables keep track of information in a way that helps you see the pattern. If Kim had studied the pattern, she might have noticed that for every dollar of the customer's bill, she could expect to receive 25 cents in tips.

Today you will work on a problem where keeping track of the information in a table will help you solve a problem and make a prediction.

Activity: Banquet Tables

Allow about 30–45 minutes for this activity. Have available square tiles (or some other manipulatives like square sticky notes) so students can visualize building the banquet tables. On the board or using a projected presentation, set up the tiles as they are on the student handout.

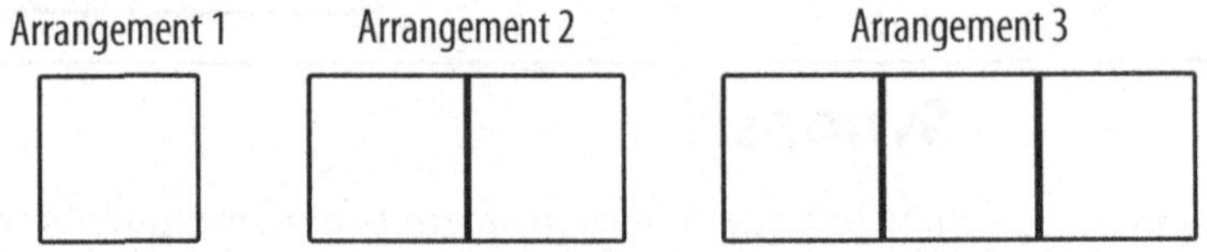

Ask:

What do you notice about this pattern?

Listen to students' comments. Then ask them to turn to *Banquet Tables* (*Student Book*, pp. 22–25), and say:

Think of yourself as the captain of the banquet room at a fancy hotel. You can arrange banquet tables to almost any length to seat almost any number of people. The table arrangements you see form a pattern that can help you predict how many tables you will need for any number of people or, conversely, how many people you can seat when you know how many tables there are.

The first picture is a small square table that seats four people. (Draw a mark to show where each person would sit.)

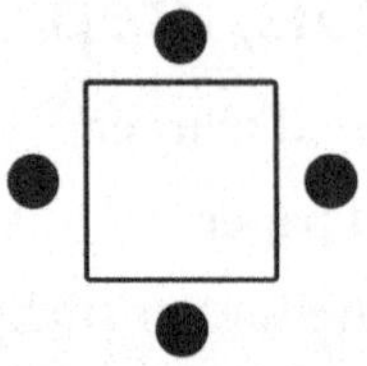

The second picture shows two of the squares pushed together to make a bigger table, and in the third arrangement three tables are pushed together. How many people would fit around the tables in each of those arrangements?

Ask students to work on the activity in pairs or individually for about 15–20 minutes. Encourage all representations—diagrams, tables, rules in words, and equations.

When most people have completed the problem, call the class together. Focus on the diagrams first (Problems 1–4).

Next, focus on the table of numbers as a way to organize data. A volunteer shares his or her table from Problem 5. Students compare the tables they made with the volunteer's, looking for similarities and differences. Column labels are important, but specific descriptions may vary. Everyone should understand that one column refers to the number of square tables and the other to the number of people. Numbers, however, should follow the rule the student(s) created.

Continue working with data from the table:

How did extending the table help you predict for different seating arrangements?

Choose a volunteer to extend the table. Check to see whether others agree or whether they wish to enter additional data. You might want to challenge the class to extend the table to include some missing data, such as 400, 401, or *n* tables, and 400 people.

Invite students to write the rules they found to determine the number of people who can sit at a given number of tables by focusing on the new entries.

Then do the same for rules to determine the number of tables for given numbers of people. One at a time, pose questions such as

How many people can be seated at the 400th table arrangement? How do you know?

How many tables will it take to seat 400 people? How do you know?

Heads Up!

Fully process each question separately—first figuring out the number of people, given the number of tables, and then figuring out the number of tables, given the number of people.

Record the methods students used to figure out the problem. Some will refer to their diagrams, others to the table. Either way, they must make the relationship explicit. Establish which letters will stand for the number of people and for the number of tables. They might be *p* and *t*. As students mention their rules, post them near one another for comparison. The board might look something like:

Table:

Number of tables (t)	Number of people (p)
1	4
2	6
3	8
4	10
10	22
100	202
9	20
49	100
400	?
401	?
?	400

Rule for finding number of people:

Adrianna's rule:
To get the number of people, add the number of tables to itself and add one on each end.

Jose's rule:
Add 1 to the number of small tables and then double that number.

Kayla's rule:
2T + 1 + 1

Marta's rule:
2(T + 1)

Darrell's rule:
2xt + 2 = p

Ask students to demonstrate how the diagrams, tables, and their rules relate. Pose questions such as

If Darrell says $p = 2 \times t + 2$, where is the $2 \times t$? Where is the + 2? In the diagram? In the table?

Kayla, where do you see the +1 + 1?

Marta's rule, $2(t + 1)$, and Darrell's, $2 \times t + 2$, look different. Are they related? If so, how?

Ask those who do not volunteer their own rules to verify the accuracy of the ones others posted by substituting in values from the table.

How can you check to make sure this rule works?

Repeat this process for the second situation—the number of tables needed to seat 400 people. Ask for examples of rules.

The board might look something like this:

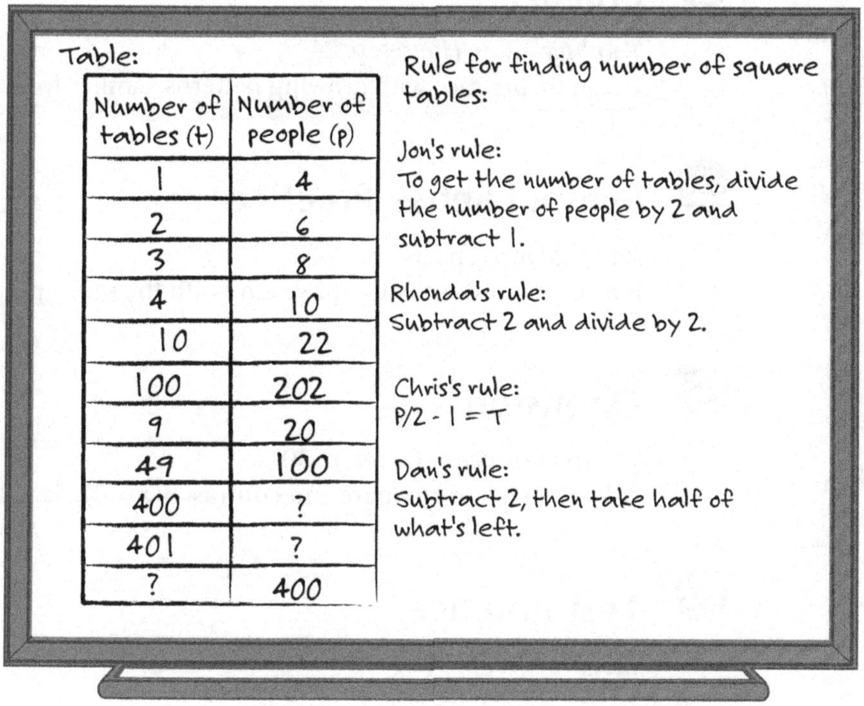

Help students process their reasoning by asking questions such as

How do Chris's and Dan's rules relate to the diagrams? Where is the division by 2 and subtraction of 1? Whose rule has the subtraction of 2 and the half?

After rules are verified, ask:

What did you learn by working on this problem?

What are you still wondering about?

Write down a few responses before moving on to the summary. If students are curious about the multiplication-division relationship, provide a few arithmetic examples to demonstrate the inverse nature of the operations.

Summary Discussion

This is a good time to revisit the definitions of "pattern" and "table" posted on the class vocabulary list. Students have more experience now and may want to refine earlier statements.

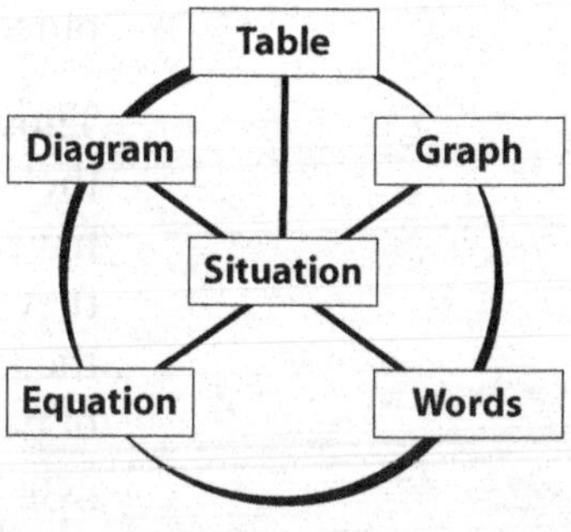

End the lesson by asking:

How did the diagram help you discover the rule?

How did the table help you discover the rule?

Practice

Toothpick Row Houses, p. 26
For more practice with growing patterns, similar to the banquet tables problem.

Symbol Sense Practice

Rules of Order, p. 28
For practice evaluating expressions with the four operations and parentheses.

Extension

The Importance of Order, p. 30
Asks students to compare and contrast $5(x + 2)$, $5x + 10$, and $5x + 2$, using tables.

Test Practice

Test Practice, p. 31

Looking Closely

Observe whether students are able to

Use objects, diagrams, tables, and rules to understand and describe a pattern in a situation

Do some students build arrangements one at a time to visualize or concretely experience the problem? Encourage them to use their diagrams or models to see the pattern. Ask questions such as "How many more people do you seat when you add another table? What happens to the two at the ends?" For these students, a table of numbers is a useful bridge between the visual and the numerical. Encourage them to extend the pattern visually, then record results in the table.

Do not press for a rule until students can readily predict how many people will be in the next arrangement and can skip a few table entries to predict further outcomes beyond those pictured.

Those who are able to generalize may do so in a variety of ways. Some explain their rule in terms of a specific case with specific numbers. These students often have difficulty writing an algebraic equation to represent the general pattern they see. Offer other numbers of tables, such as 12 or 17, and ask what happens then. Have the students dictate to you their process for figuring out the number of people who can sit at those tables. Use their descriptions to model a more general statement, substituting words like "any number of tables—*n*" for specific numbers.

Do some students state a general rule in words, but not symbolically? The potential to move quickly to symbolic representation likely relies on the verbal clarity expressed. Offer students a stem to start with: "To find the number of people, take the number of tables and..."

Some students use only symbols to generalize, indicating a preference for the algebraic shorthand over verbalization. This can frequently be the case with English for Speakers of Other Languages (ESOL) students who are often proficient mathematically, but struggle with English. Some students are able to express rules fully in words and accurately in symbols. Sharing the verbal rules of these students with the class will help all who struggle to articulate rules. Having students who can write symbolic rules connect their rules with concrete examples helps those who have difficulty generalizing from specifics.

Over-application of proportion can be a stumbling block for some students. Watch to see whether they are applying a workable rule, or simply multiplying by 10 when moving from the 10th to the 100th case.

See *Lesson 2 in Action*, p. 38, for specific examples of student work representing these various approaches.

Keep track of problem-related data in a table

Do some people make a table, then forget to use it? Remind them that they can keep track of all information in a table, not just that requested in the problem. Also remind them how they used the tables in *Guess My Rule* to figure out a rule that let them predict an outcome for any number. Would a table here help in the same way?

WHAT TO LOOK FOR IN *LESSON 2*	WHO STANDS OUT? (LIST STUDENTS' INITIALS)			NOTES FOR NEXT STEPS
	STRONG	ADEQUATE	NEEDS WORK	
Representations Diagrams • Accurately diagrams Banquet Tables Arrangement 4 • Transitions from creating diagrams of table arrangements to applying pattern to 100th arrangement Tables • Constructs a table with appropriate column headings • Identifies relationship between *x*- and *y*-values within the table • Predicts missing values correctly Rules • Accurately describes relationship between *x*- and *y*-values in words • Accurately writes relationship between *x*- and *y*-values as algebraic equations • Uses algebraic notation conventions for all four mathematical operations				
Connections • Uses table data to arrive at rule in words • Connects diagrams with features of the equations				

Rationale

Once students have gained some familiarity with searching for and describing patterns in tables, they benefit from exposure to situations in which a table helps them solve a problem. This way, they come to understand that there are many methods, beyond arithmetic, to approach a new problem.

Math Background

"Growth problems" can be used to demonstrate a variety of functions. *Banquet Tables* represents a linear function in which $y = 2(x + 1)$. The advantage of these types of problems is that they allow students to connect the algebraic or symbolic rule with the visual or concrete representations of the arrangements. By building the pattern, they can see where the **coefficient** 2 in $2x$ comes from (the number of people added on either side each time a table is added) and where the **constant,** +2, comes from (the two seats at the ends of each table arrangement).

Later, when students graph the *Banquet Tables* function in *Lesson 4*, they will see the coefficient and the constant in another way.

The inverse relationship between multiplication and division surfaces in this problem. You multiply to find the number of people and divide to find the number of tables. Working through equations to demonstrate this relationship algebraically may be too advanced for most students. Provide arithmetic examples as a reminder that $2 \times 4 = 8$ and $8 \div 2 = 4$, and so on.

Facilitation

Pairing Students: *Banquet Tables* is a problem with many points of entry, from the symbolic to the concrete. When pairing students, avoid having an algebraically minded student frustrated by one who needs to draw each example, and vice versa. Encourage students to learn from each other by comparing their representations.

Debriefing: Processing the two different rules—from table to people and from people to tables—requires space. Recording students' rules on easel paper and posting these sheets for all to see allows students to compare the rules easily.

Making the Lesson Easier

Do not emphasize the algebraic notation.

Making the Lesson Harder

Connect the equivalent algebraic expressions. For example, what does $p = 2t + 2$ have to do with $t + t + 2 = p$ or with $2(t - 2) + 2(3)$?

Explore the inverse relationship of the two rules ($p = 2t + 2$ and $t = p/2 - 1$ in greater detail.

Students' diagrams often gave teachers an indication of how quickly they were able to generalize.

By the 100th example, some people continued to draw every table in a literal graphic representation of the problem.

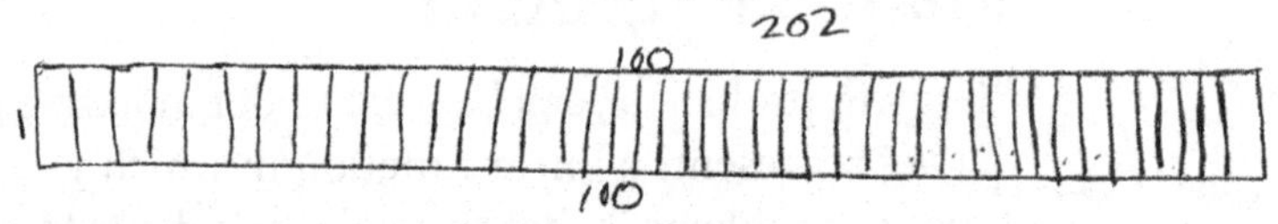

Others produced a more abstract picture.

However, some students were able to utilize symbols by the time they were thinking about the 100th arrangement of tables.

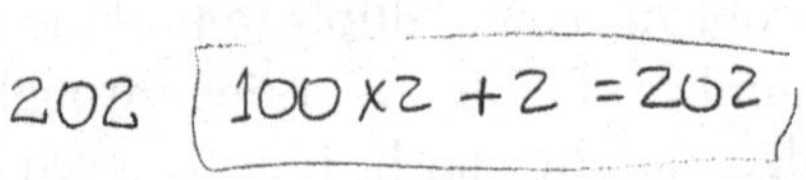

First attempts to formulate rules also revealed various levels of generalization.

Some students held on to specific numbers.

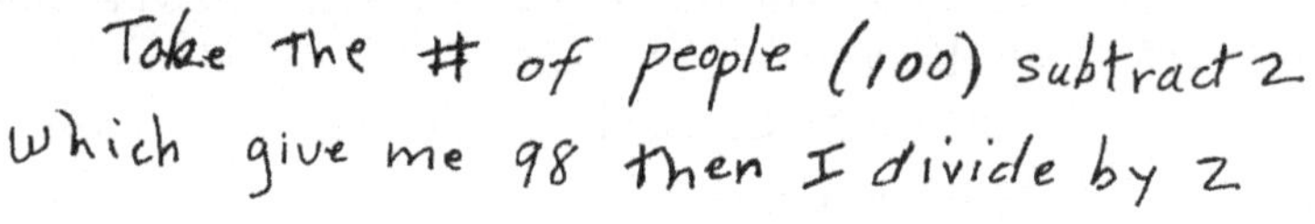

Some began to communicate a more generalized view of the pattern that would work regardless of the numbers involved.

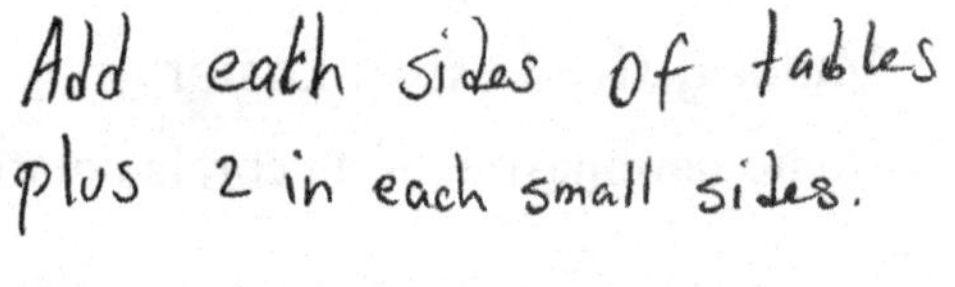

continued on next page

continued from previous page

Others wrote an algebraic rule as well as a verbal rule.

$\frac{\text{People}}{2} - 1 = \text{Tables}$

Divide the amount of people by two, then take the answer and minus one, to get the number of tables.

Students used several ways to visually connect the diagrams to the rules.

Three rules were explained as

$p = 2t + 2$. Some students noticed that there were two people across from each other at each table ($2t$) and 2 people on the ends ($+ 2$).

$p = 2(t + 1)$. Others saw that they could add the person on one end to the number of people along one side ($t + 1$), and then double that amount to get the answer.

$p = 2(t - 2) + 2(3)$ or $2(t - 2) + 6$. Some separated the people on the two end tables (6) and then doubled the number of tables that remained $2(t - 2)$.

All of these students participated actively in discussions. The challenge was to connect each one's work with the others'.

Mary Jane Schmitt
Bridge to Learning and Literacy Program,
Harvard University, Cambridge, MA

FACILITATING LESSON **3**

Body at Work—Tables and Rules

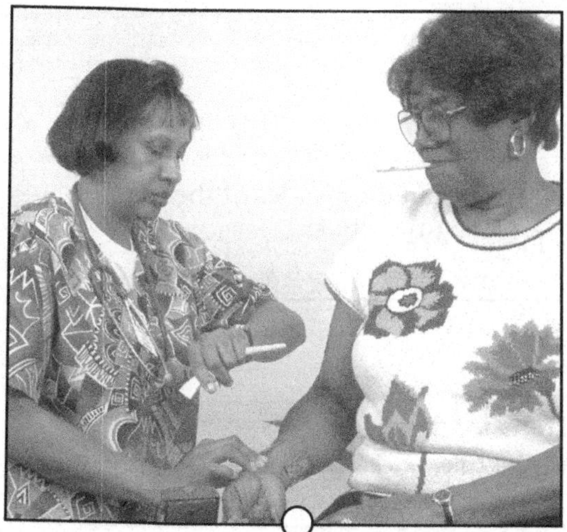

What is your heart rate at rest?

Synopsis

This is the first of a set of three lessons in which students explore, extend, and compare patterns about the human body. This lesson emphasizes connecting rules with tables of real data, and using tables and rules to solve related problems.

1. Students take their pulses at rest for 15 seconds, make a class table of the data, and state a rule for finding the number of beats per minute.
2. Individuals develop a personal table, a general statement, and an algebraic equation to determine the number of times their own hearts beat in any number of minutes.
3. Student pairs complete four tables with information about the number of calories burned over time for different exercises. They write a rule in words and algebraic symbols for each table.
4. Summary discussion centers on strategies for relating tables and rules, and using the table or rule to help solve problems.

Objectives

- Describe the pattern in a situation with a verbal and symbolic rule
- Connect patterns in tables with generalized rules
- Solve problems using the patterns represented in table data and rules

Materials/Prep

- Calculators
- Clock or watches with second hands
- Colored markers
- Graph paper
- Easel paper
- Rulers

For *Activity 2*, prepare easel paper copies or presentation slides of the four "Calories Burned" tables (*Student Book*, pp. 37–39).

Heads Up!

Use the board or set up easel paper copies of the data for the whole class to see so they can compare the various representations (table, verbal rule, and algebraic equation).

Opening Discussion

Explain that this lesson again focuses on noticing patterns and stating a rule to solve a problem, but that this time the data concern students themselves.

Begin with a common pattern for everyone—heart rate, or pulse. Say:

Start by thinking about a pattern that everyone has—a heart rate, or pulse.

How does a nurse or a workout trainer usually take a pulse?

Students will most likely mention that nurses and trainers check pulse rates on the wrist. If no one mentions that a pulse is usually taken for 10 or 15 seconds, explain that this method saves time.

Activity 1: Heart Rates at Rest

Part 1

Ask everyone to take his or her pulse for 15 seconds, and post the results in a table for all to see. The results might look something like this:

Name	Beats per 15 Sec.	Beats per Minute
Jonna	20	
Ron	17	
Denise	18	

Then ask:

How can you predict the number of times your heart will beat in a minute?

After students copy the information in the table onto their charts (*Student Book*, p. 34), they are ready to fill in their predictions for the pulse rate for a baby and a feverish adult.

Ask:

How did you figure out those two new listings? Did anyone do it a different way?

What is the rule for finding the total number of heartbeats in one minute?

Write down people's *verbal* rules verbatim on the board. In one class, for example, students dictated:

Rule in Words

To find the heartbeats in a minute, multiply the heartbeats in 15 seconds by 4.

Heartbeats in 15 seconds times four equals heartbeats in one minute.

Times it by two, two times.

Add the 15-second number four times.

Check for agreement on the ways the rules are expressed. While you will be working toward using precise language, use the students' words and notation for now.

Ask for volunteers to translate the verbal rules to algebraic notation. You might prompt by saying that "F" stands for the number of beats in 15 seconds and "M" for the number of beats per minute. In one class, the students wrote:

Rule in Words		Algebraic Notation
To find the heartbeats in a minute, multiply the heartbeats in 15 seconds by 4.	→	M = F x 4 M = 4F
Heartbeats in 15 seconds times four equals heartbeats in one minute.	→	F x 4 = M
Times it by two, two times.	→	F x 2 x 2
Add the 15-second number four times.	→	F + F + F + F = M

Check for agreement on the way the algebraic expressions are written, and use this opportunity to ask people to relate the various algebraic expressions:

How can you show with words or by sketching that your algebraic expressions mean the same thing?

Part 2

Students now focus on their own heart rates. They complete their personal heart rate table (*Student Book*, p. 35). Starting with their one-minute pulse rates, they determine the number of times their hearts beat in other time periods, such as in 5 minutes, 10 minutes, an hour (60 minutes), or a day (1,440 minutes). Ask student pairs to compare tables with each other. When everyone has a personal table completed, suggest:

Think of the table you completed as an In-Out table. Is there a rule you can write that tells how you get from the number of minutes to the total number of heartbeats?

Call upon three volunteers to record their tables and rules in words and in algebraic notation on the board. Focus on one person at a time. Draw attention to the connections between each table, its verbal rule, and its algebraic rule, asking:

Show how your rule and table are related.

Does the rule work for every instance in the table?

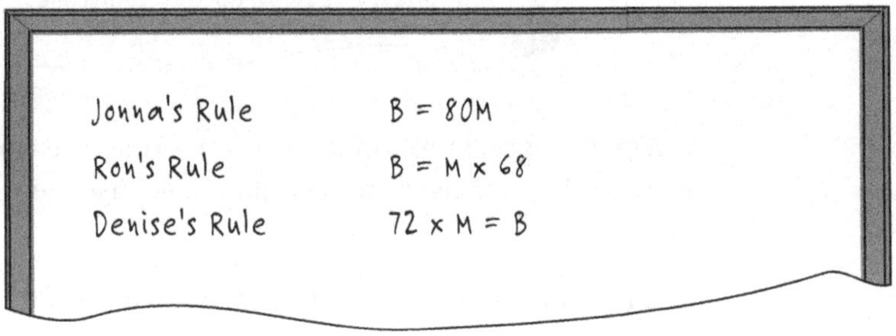

Students should use table data to verify that a rule works.

Select a fourth student's algebraic equation to write on the board, and pose the question:

Whose rule might this be? How do you know?

Then instruct:

Use the rule to predict the number of times this person's heart will beat in a day.

Ask the originator of the rule to verify the prediction.

Activity 2: How Many Calories Am I Burning?

Explain that the next activity will provide a chance to think more about how bodies work and how patterns and rules can be used to predict future outcomes or solve problems.

Form pairs or small groups of students. Direct attention to *How Many Calories Am I Burning?* (*Student Book*, p. 37). Let pairs work for a while, and then ask for volunteers to fill in the easel paper tables you prepared earlier.

How did you know which numbers to use to fill in the tables?

What pattern did you see in Table 1? Table 2? Table 3? Table 4?

How did those patterns show up in your rules?

Ask a few volunteers to write verbal and symbolic rules for each table on the easel paper and to connect those rules with the table data.

Heads Up!

The tables show that jogging burns 10 cal./min.; cleaning house burns 5 cal./min.; running up stairs burns 20 cal./min.; and watching TV burns 1 cal./min. Rule for jogging, $y = 10x$; cleaning house, $y = 5x$; sitting, $y = x$; and running up stairs, $y = 20x$.

Next, focus attention on how people used the tables and rules to solve Problem 1:

How long do you have to watch TV to burn the same number of calories as you would in a half-hour of jogging? How do you know?

If we substitute those times (30 minutes of jogging and 300 minutes of TV watching) into these rules, what will happen?

Did anyone use his or her rules to solve this problem? How?

Substitute the number of minutes for each of those activities into the symbolic rules to show that the calorie totals are equal. If no one used his or her rules to solve the problem originally, work on that now.

Summary Discussion

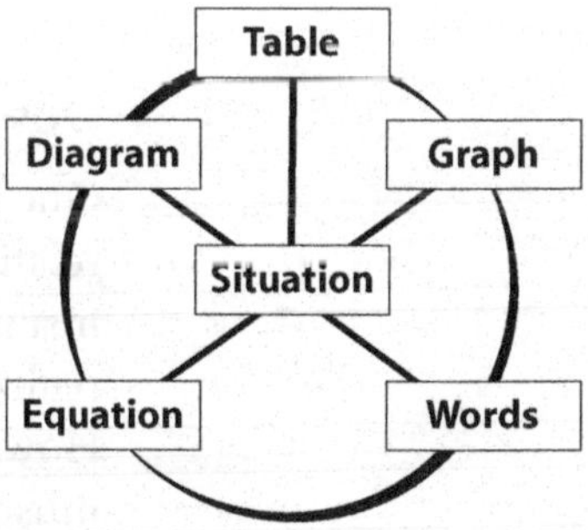

Much happens in this lesson. Take time to review what has been learned by asking:

How did you use the tables and/or algebraic equations to solve problems about heart rates and calories burned?

How can you tell that a table and a rule are related?

Also, provide an opportunity for students to share concerns and achievements by asking:

What was hard and what was easy for you today? What questions do you still have?

Finally, suggest students take a few moments to write in *Reflections* (*Student Book*, p. 159).

Practice

Say It in Words and Fill in the Tables, p. 41
For practice translating common abbreviations for rates into tables and connecting those to situations.

Driving at 50 Miles per Hour, p. 42
For practice looking at the rule in a table that involves distances and times.

Symbol Sense Practice

Equations ⟷ *Words*, p. 44
Asks students to translate algebraic expressions to words and vice versa.

Substituting for x, p. 45
For practice evaluating expressions.

Extension

A Friendly Reunion, p. 46
Extends this lesson to a more complicated situation that involves times, distances, and speed. A follow up to *Driving at 50 Miles per Hour*.

Test Practice

Test Practice, p. 48

Looking Closely

Observe whether students are able to

Describe the pattern in a situation with a verbal and symbolic rule

One issue that impedes rule description occurs when students rely on numerical relationships that surface in the columns, rather than looking across the rows. For instance, if the figure for calories burned for 15 minutes is given, students may realize they can double that figure to arrive at the calories burned in 30 minutes. However, this will not help them find the rule for this situation. To do that, they must look across the rows. Remind them that while these tables reflect particular levels of activity, they operate like In-Out tables. Pushing to the 100th case often forces the need to define the rule as well.

Work on getting the verbal rule as precise as possible, and use mathematical terms like "multiply it by 10." The *Symbol Sense Practice* (*Student Book*, p. 44), provides opportunities to translate equations into words and helps students become familiar with the language-symbol connection. More explicit work with rules and equations takes place in *Lesson 7*.

Do students find it hard or easy to work with letters representing the variables? For the most part, let students define the variables in their own ways. However, sometimes it is useful to make a suggestion. Make sure students understand that the variable represents the *number* of heartbeats or the *number* of total calories and that these numbers vary according to the particulars of the situation.

How well do students connect the verbal rule to the equation? One way to help students translate verbal rules into notation is to write out the verbal rule, formulated as specifically as possible, across the board or on a piece of paper. Then work with the student to connect the words with symbols and numbers, using arrows.

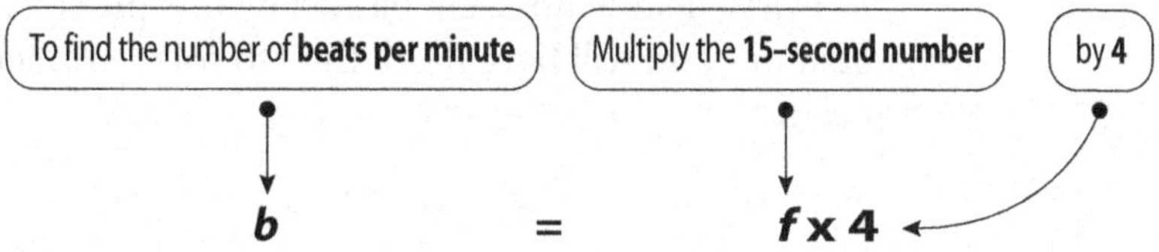

Connect patterns in tables with generalized rules

Do students see that the multiplier (the coefficient) in the equation is the number that they multiply time-related variables by to get figures for heartbeats or calories burned? Make that connection explicit by setting up a third column in the table between the time column and the heartbeats or calories-burned column. Ask students to write in this space what they did to get new numbers. Then ask:

Where is that number in your rule?

Minutes Jogging (m)	$10m = C$	Calories Burned (C)
1	x 10 =	10
15	x 10 =	150
30	x 10 =	300
45	x 10 =	450

Asking students to check rules by substituting table values into the equations or the verbal rules further strengthens the connections.

Solve problems using the patterns represented in table data and rules

Do students show how a solution is supported by the table data and the rule? Although some of the problems can be solved with simple computational arithmetic, ask students to connect this information to the table and rule as well.

Spend time showing how the table data for TV-viewing might be extended to reach 300 minutes without making every 15-minute entry. Match the calories-burned figure for TV-viewing to the jogging-for-30-minutes entry. Likewise, demonstrate how the rules could be used to solve the problem:

Jogging – $C = 10m$, so $C = 10(30)$ or $C = 300$

Watching TV (sitting) $C = 1m$. We know we want to burn 300 calories, so $C = 300$, and $300 = 1(m)$. What number multiplied by 1 equals 300?

Always use students' rules, and ask students to demonstrate how they might use them to solve the problem first. If they do not know how, then demonstrate. Do not worry if students seem baffled by this process. They are just getting used to equations and will have more time to solve problems with them in the lessons ahead.

WHAT TO LOOK FOR IN *LESSON 3*	WHO STANDS OUT? (LIST STUDENTS' INITIALS)			NOTES FOR NEXT STEPS
	STRONG	ADEQUATE	NEEDS WORK	
Representations Diagrams • Creates a diagram to represent the number of heartbeats in 15 seconds related to the number of beats in a minute Tables • Recognizes vertical and horizontal patterns • Identifies relationship between *x*- and *y*-values • Predicts missing values correctly Rules • Accurately describes relationship between *x*- and *y*-values in words • Accurately writes relationship between *x*- and *y*-values as algebraic equations • Uses algebraic notation conventions for all four operations				
Connections • Uses table data to arrive at rule in words • Recognizes how table data connect to symbolic rules				

Rationale

Tables and rules continue to be the focus in this lesson, but students extend their knowledge by connecting tables and rules to real situations. Connecting tables to rules and using each to solve problems involving bodily data demonstrates that the pattern description skills learned earlier can prove meaningful in our daily lives.

Math Background

Rates: References to rates are endemic to our culture. We talk of miles per gallon, miles per hour, interest rates, heart rates, etc. all the time. However, understanding the mathematical relationships involved in rates is often difficult. Most commonly used rates involve change over time or distance. The dependence of one variable on the other—of miles per gallon of gas or interest paid for a number of months or years—can be confusing to anyone whose sense of proportion is shaky. This is because rates involve two variables, either of which can influence the magnitude of the other. As proportional relationships, common rates can be represented as linear relationships on graphs, which will be dealt with in the next lesson. Proportional relationships are not the main focus of this lesson, but inform much of the work.

Symbol Sense Focus: How much you decide to focus on symbol sense depends on the needs of the class. During the lesson, you might choose to highlight any or all of the following:

- Ways to write about rate: 60 mph, 60 mi./hr., 60 miles per hour, 60 miles in 1 hour.
- Ways that rates can look different when extended but be constant: 60 beats in 1 min. is the same rate as 600 beats in 10 min.; 60 beats/min. = 600 beats/10 min.
- Ways to write multiplication without a multiplication sign:

  $C = 5m$ reads C equals 5 multiplied by m; $5m = 5(m)$.

- Relationships that use multiplication to find a total—for instance, total calories burned—can be reformulated as division rules to find one of the units multiplied, e.g., minutes exercised:

  $C = m(10)$ for dancing and $\frac{C}{10} = m$ or $\frac{C}{m} = 10$

Context

Heart Rates: Differences in heart rates interest students. Normally the heart beats 60–80 beats per minute (bpm), although it can beat up to 200 or more bpm during intense exercise periods. Exercise is not the only stimulus that can raise a person's heart rate, however. Drugs, such as caffeine and nicotine; hormones, like epinephrine and those produced by the thyroid; and mental conditions, like anxiety, can all raise a person's heart rate. High body temperatures, like those experienced during a fever, can also increase the heart rate to make you feel like your heart is racing. Low body temperatures decrease the heart rate, as does being in good physical condition.

Calories: A calorie is a unit of energy equal to the amount of heat energy it takes to raise the temperature of one gram of water one degree Celsius. The calorie amounts listed on food packages are actually kilocalories. When a package reads that two slices of bread have 150 calories, it means "food" calories, or kilocalories. In scientific terms, it really has 1,000 times that many calories, or 150,000.

For more information, go to https://health.howstuffworks.com/wellness/diet-fitness/weight-loss/calorie.htm.

At https://www.healthstatus.com/calculate/cbc/ you can calculate calories burned for any of dozens of activities. The site charts the calories burned based on height, weight, gender, age, etc.

At http://www.nutristrategy.com you can find tables of calories burned for various activities.

Facilitation

Numeracy Connection: You and your students might look for tables outside the classroom, perhaps sent with a bill, in an advertisement, or in the newspaper. For each table, you might ask:

What patterns do you notice?

Does there seem to be a rule?

Is there a rule to predict what comes next in the table?

Students use a table to organize heartbeats in 15 seconds and the corresponding pulse in a minute. Make sure everyone reads the column headings and understands what they mean.

Some students have trouble knowing where to look for the pattern. Remind them that these tables are similar to In-Out tables.

Making the Lesson Easier

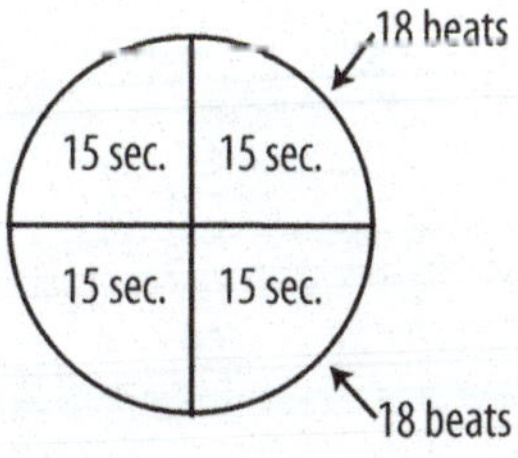

Conduct the two activities as separate lessons. Spend more time on the heart rates exercise and the relationship between the number of beats and number of seconds.

Using objects to model these relationships can help demonstrate their equivalency. Ask students to diagram or represent the relationship between time and beats by having pennies represent seconds and large paper clips represent heartbeats. Someone can lay out a penny for each of 15 seconds while another person takes his or her pulse. Then lay out the paper clips for the heartbeats. Look to see how the clips and pennies relate. Is there is a one-to-one correspondence? Ask students how many paper clips or heartbeats there are for each second. Place the pennies in an arc from 12 o'clock to 3 o'clock, and ask students how many groups of 15 pennies would be needed to fill a whole clock. Ask them to figure out how many heartbeats would then be represented.

Making the Lesson Harder

If most students are comfortable with algebraic notation, spend a little more time being more precise about the notation, comparing and relating the multiplication form with the division.

If $C = 50m$, then $m = \frac{C}{50}$.

Ask students:

Why do we multiply to find the number of calories burned, but divide to find the number of minutes doing the exercise?

LESSON 3 IN ACTION

The unit's early lessons are launched with a situation itself (e.g., taking pulses) and two-variable tables of numbers. Students are asked to seek patterns and to write a rule that explains how to find the second variable, when the first is known. They are strongly encouraged to say or write the rule in everyday language before they write it in algebraic symbols because, as one student we videotaped said, "I understand it more when I try to write it down in words."

Pulse Rate Activity

When we look closely at how students verbally express rules, we see a variety of responses. Most responses have in common the communication of a mathematical pattern, but they vary in degree of precision. Here are examples from pilot classes.

Some of the language is very precise:

"The nurse takes the pulse for 15 seconds and whatever number she gets, she times it by four."

"The pulse rate for 15 sec. x 4."

"Take pulse for 15 seconds and multiply by 4."

"Multiplies beats for 15 sec. by 4."

"Multiply beats/15 seconds by 4."

Some of the language is less complete and less precise, but you can understand what the student means.

"Multiply by 4" (several students wrote this).

"x4."

"She multiplies 15 seconds by 4 to get the answer."

"Multiply by pulse rate x 4 = ____."

"Multiply the number of ♫'s by 4."

"Times by 4."

"15 seconds x 4 = _____."

"Multiply the beats by 4."

Some students did not generalize:

"18 beat to 15 sec."

Ask yourself: What kind of instruction will move students toward more precise language?

continued on next page

continued from previous page

In one class with both General Educational Development (GED) and pre-GED students, there was one student who did not know that there are 60 seconds in a minute. He guessed there were 100. Not knowing this basic fact made it hard for him to grasp the multiplication in the opening heart-rate exercise. It was not clear to him that you would multiply the number of beats in 15 seconds by four because 4 x 15 did not equal 100. This young man wrote at the end of class that he enjoyed working on the problems, but experienced some confusion when doing problems with heartbeats and minutes. In fact, he posted his heart rate as 15 beats in 15 seconds, which made us wonder if he counted seconds, not heartbeats. He couldn't distinguish between the two variables and see them as distinct, yet related.

Calories Burned Activity

When working on the calorie charts, the level of computational ability separated those who completed the table quickly from those who took more time. Fortunately, people helped one another.

In a pre-GED class with a wide skill range, some people divided and multiplied with ease; others figured out situations calling for multiplication and division by using adding or doubling techniques. Even with calculators available, the computation was a workout for some.

At the close of one class, students were asked how the class went for them. One student said, "I learned more about burning calories. It went fast and was fun."

Compilation of data from several pilot classes

Bridge to Learning and Literacy Program,
Harvard University, Cambridge, MA

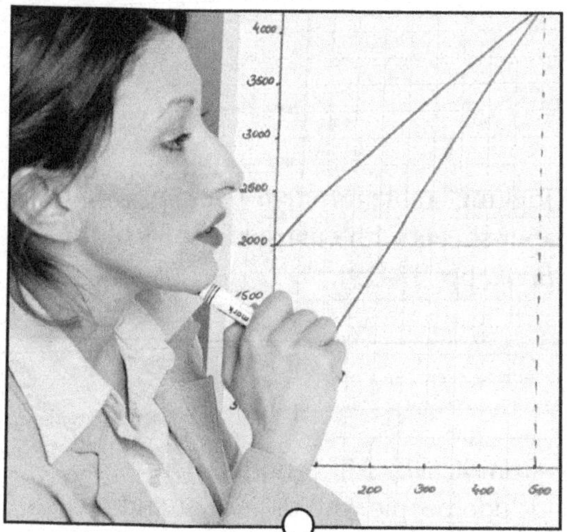

FACILITATING LESSON 4

Body at Work—Graphing the Information

Where is the pattern in the graph?

Synopsis

This lesson introduces graphing as students continue to work with the "Burning Calories" tables from *Lesson 3*. Students make connections between various exercise graphs, as well as between the rules and tables they represent.

1. Drawing upon their collective knowledge of graphing conventions, the whole class constructs a graph from the information in the "Calories Burned Jogging" table.
2. Student pairs construct this same graph on grid paper and add the graphs from three other activities.
3. The class compares the graphs' features.
4. Students connect graph features with the number patterns in the tables and the rules.

Objectives

- Identify and use features of graphs (origin, *x*-axis, *y*-axis, scale increments, points as **ordered pairs** of **coordinates**, slant of the line)
- Use information from tables or rules to generate graphs
- Connect tables and rules with their corresponding graphs

Materials/Prep

- Calculators
- Colored markers
- Graph paper
- Easel paper
- Rulers
- Yardsticks (two)

Prepare a copy of the "Calories Burned Jogging" table (*Student Book*, p. 37), to post for the whole class. Prepare easel paper sheets with enlarged copies of the four activity tables in *Lesson 3* (*Student Book*, pp. 37–39).

Heads Up!

Everyone will need the completed tables and rules from *Activity 2* in *Lesson 3, How Many Calories Am I Burning?* Have a plan ready to include people who did not attend the last class.

A note on vocabulary: The **x-axis** is the horizontal line on a graph. The **y-axis** is the vertical line on a graph. Each axis is divided into evenly spaced parts, called **increments.** A **point** on a graph is labeled with an x-value and a y-value. For example, the point (2, 5) indicates the x-value of 2 and the y-value of 5. The **origin** is the point where the x- and y-axes meet. It is labeled (0, 0).

Opening Discussion

Review the tables and rules discussed in the last lesson. Tell the class that algebra uses graphs, as well as tables and rules, to describe patterns. Say something like:

Graphs give a new look to the patterns described in tables and rules. By the end of class, you will have a much clearer sense of how a pattern that shows a constant rate can be represented in several ways—in a table, as a rule, or on a graph.

Activity 1: Four Graphs for Calories Burned

Building a Whole-Class Graph

Guide the whole class in constructing a graph of the data on burning calories from the "Calories Burned Jogging" table.

Ask volunteers to share what they know about making graphs. Have different people set up and label the axes and insert the increments on them. Increments are often especially tricky; take time to demonstrate why equal spacing is important and the need for planning the size of the increments ahead of time so the numbers fit on the axes.

Then plot at least three points from the "Calories Burned Jogging" table (use two yardsticks to demonstrate plotting), and draw the line.

Check to see who remembers the convention for **labeling points**, and ask these students to demonstrate by labeling several. If no one remembers how to label graph points, write the ordered pair (1, 10) next to that point on the graph. Ask where the "1" and the "10" come from. Volunteers can label more points and explain where each value comes from.

Emphasize that points are always labeled in alphabetical order, with the x number first and the y number second.

Discuss how graph increments normally start at zero and increase, moving to the right on the x-axis and upward on the y-axis. Practice finding a few points *between* the increments.

The class's graph will probably look something like this:

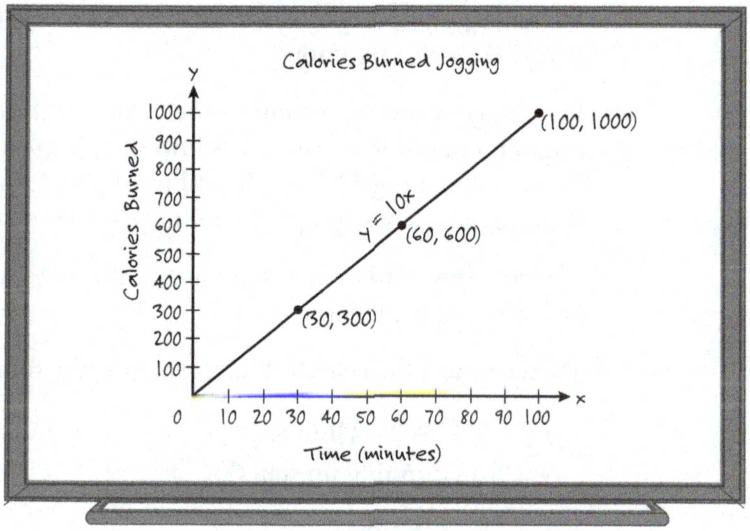

Review vocabulary. Ask volunteers to describe the terms ***x*-axis, *y*-axis, increments, labeling points,** and **origin,** so these can be posted on the class vocabulary list. Frame questions such as

What does the word "origin" bring to mind?

Then connect the idea of "beginnings" with the origin of the axes. When the graph is complete, ask:

Looking at the graph, what do you notice?

Who can summarize for us how the plotted points connect to the numbers in the "Calories Burned Jogging" table ?

Sharing observations about the graph shows you what features students are noticing. It is a good time to clarify misconceptions, as well. When everyone seems comfortable with coordinate graphing conventions, ask students to reproduce the "Calories Burned Jogging" graph in their books on p. 53.

With partners, students proceed to make graphs for the other three exercises. All four graphs should be created on one grid. When most students have constructed all four graphs, post all or a few, and ask students to compare them:

What do you notice when you compare our work?

What is different? What is the same?

If students use different increments for the scale on their axes, the graphs will appear different. You might point out the connection between increment choices and the look of the graph, but be sure to focus on the *direction* of the line's slant—up to the right. Students need to see that regardless of the type of exercise, when calories increase as time increases, the linear graphs always slant in this direction.

Choose one student pair's work to focus on, and ask:

From a distance, can you tell which line graph tells the "Calories Burned Watching TV" story? Which tells the "Calories Burned Running Up Stairs" story? How do you know?

Encourage students to connect their statements to features of the graphs. You might point out that everyone's lowest, or flattest, line seems to be the "Calories Burned Watching TV" graph, and most people have the "Calories Burned Running Up Stairs" graph as the steepest line. Ask why this is so.

Mention the word "slope" here in an informal way. Do not press for a formal definition of the term.

Then connect the rules to the graph. Post the four algebraic rules below:

$y = 5x$ (Cleaning house)

$y = 20x$ (Running up stairs)

$y = 10x$ (Jogging)

$y = x$ or $y = 1x$ (Watching TV)

Ask:

How do you know which rule goes with which graph?

Students might say something like, "This graph goes up by 10's or 5's, and so does the rule." Work on these connections, and develop the notion that the line gets steeper as the number of calories burned by an exercise increases or as the speed of the activity increases. Ask students to apply their understanding with a question such as

In an exercise that has a rule where the calories burned equals 12 times the number of minutes exercised, or $y = 12x$, what do you think the line would look like? Why?

Offer several other hypothetical rules until students clearly see the connection between the rule's multiplier (x-coefficient) and the relative steepness of the lines. Students should be encouraged to come to the board to relate the labeled points on their graphs with the table entries.

Summary Discussion

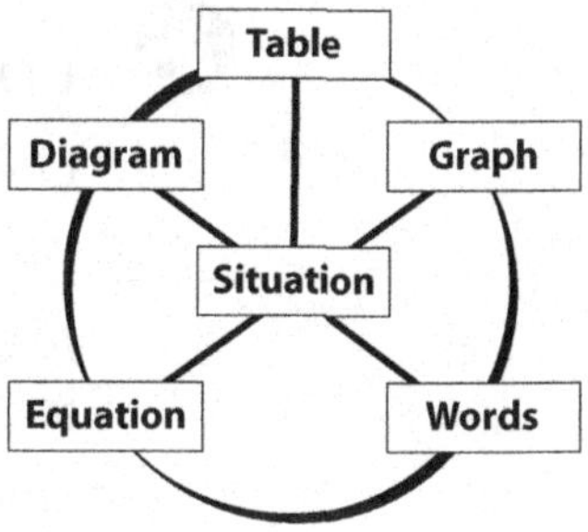

Tell the class that in this lesson they explored several important ideas in algebra. They saw that patterns can be represented as straight-line graphs, and they saw how those graphs are connected to rules and tables with the same patterns.

Next, tell the class:

As a way to review all you learned today, I would like to go around the room and ask each of you for one idea you want to remember.

Allow everyone to speak, even if there is repetition. You might prompt for new information if people get stuck by asking about graph terms like x-axis and y-axis and about the connections between graphs, rules, and tables.

To reinforce what was learned, you might ask students to make a Mind Map of the word "graph" in their *Reflections* section (*Student Book*, p. 159).

Practice

Setting Up A Coordinate Graph, p. 56
For a review of the vocabulary introduced in the lesson.

Graphing Guess My Rule, p. 57
An opportunity for students to draw and compare graphs for three In-Out tables from *Lesson 1*.

Graphing Banquet Tables, p. 60
An opportunity for students to represent the pattern from *Lesson 2* in a graph.

Symbol Sense Practice

Relating Multiplication and Division, p. 61
Asks students to change multiplication equations to corresponding division equations.

Test Practice

Test Practice, p. 63

Looking Closely

Observe whether students are able to

Identify and use features of graphs (origin, *x*-axis, *y*-axis, scale increments, points as ordered pairs, slant of the line)

Students who have difficulty graphing may simply need more practice; however, it is important to notice whether the difference between the *x*- and the *y*-axes is clear.

Remind students that they need to recheck the table values and the placement of points in the coordinate plane. Using a ruler to keep track of graphing points may help. Also, making sure points are labeled confirms table data and allows easy reference to the axes.

Use information from tables or rules to generate graphs

More advanced students may be able to use the rules to generate points on the graph, but most students will work directly from their tables to find *x* and *y* values. This is mainly a tracking task; students must be able to transfer information from one form to another without losing track. Those students experiencing difficulties may find it useful to mark the tables with an "*x*" and a "*y*" above the "Time" and "Calories Burned" columns.

Connect tables and rules with their corresponding graphs

Labeling points is crucial to understanding the connection between tables and graphs. Ask students:

Where is this point in your table? (Choose one that is labeled correctly.)

Follow up by asking what the next entry on the table would be and where that point would fit on the graph.

Rules are often more difficult to connect to graphs. Connect the point (1, 10), for example, to the rule $y = 10x$ and then ask:

To get to the next point on the graph, by how many increments do you move to the right on the *x*-axis; by how many do you go up on the *y*-axis?

Have students trace the steps and repeat the words "Over___, Up___," drawing attention to the repetition. Then ask how this repeated phrase relates to the rule.

WHAT TO LOOK FOR IN *LESSON 4*	WHO STANDS OUT? (LIST STUDENTS' INITIALS)			NOTES FOR NEXT STEPS
	STRONG	ADEQUATE	NEEDS WORK	
Representations Graphs • Labels *x*- and *y*-axes accurately • Chooses appropriate increments • Labels increments consistently and accurately • Labels points accurately • Uses algebraic notation conventions for all four mathematical operations				
Connections • Uses table data to accurately graph points • Uses graph points to create table data • Given several rules in the form ($y = mx$), predicts and compares graph features with table data				

Rationale

Coordinate graphing is essential at all levels of mathematics. Graphs are encountered daily by most students, so understanding how they are constructed and how they are related to table data is important. Graphs also provide students with a concrete image of a pattern. With a graph, they can see what happens to numbers and, therefore, what happens in a given situation—the line gets steeper as the change accelerates and flattens as the change slows.

Math Background

Coordinate graphs are visual models of mathematical relationships. With linear functions (the focus of this unit), the graph always forms a straight line, with one y-value for every x-value. The line may slope either up to the right (positive) or down to the right (negative), depending on what happens to y as x increases.

Although it is easiest for novice graph-makers to plot the points of the graph from table data, it is important to remember that verbal and algebraic rules can be used to generate graph points. For instance, when the rule is $y = 5x$, you can pick any point on the x-axis and multiply it by 5 to find the corresponding y-value.

A common graphing convention is that the **independent** (known) **variable** is assigned to the x-axis, and the **dependent variable** (the one affected) to the y-axis. The number of calories burned depends on the amount of time spent on an activity. In fact, many functions depend on time, so time is often assigned to the x-axis.

The slope of a line is introduced intuitively here. The concept of slope as delta y over delta x (change in y over change in x) or the concept of the line as a specific example of $y = mx + b$, where m is the slope, is beyond the scope of this lesson. Slope is addressed further in later lessons.

Facilitation

Starting the Activity

The class may exhibit a wide range of familiarity with graphing; some students may have had a lot of experience with graphing, and others may not. In the opening activity, draw on students' prior experience with graphs by asking all to work together at first, consulting with one another on how to make the graph as accurate as possible.

Increments

Making decisions about increment spacing is notoriously difficult for many students. It requires planning ahead, and many times students need to be reminded to think about how they will fit all the numbers on their axes. They need to know, too, that increments can be spaced by any multiple—1's, 5's, 10's,

100's, etc.—but that the increments must be spaced and labeled consistently. Some students will need a lot of practice with this.

Making the Lesson Easier

Have ready a copy of a line graph for students to examine and label with mathematical terms: x-axis, y-axis, and origin.

Making the Lesson Harder

Push for precision in language. Ask more "what if" questions about changing the slope—e.g., "What if the rule were $y = \frac{1}{2}x$ or $y = 2.5x$?"

The student work in Figure 1 shows that the student understands the domain (x-values) and range (y-values) and can produce a graph from the "Calories Burned" tables. She also marked the axes in equal and appropriately sized increments.

One challenge that surfaced frequently in this introduction to graphing had to do with determining the size and spacing of the increments on the axes. For example, in Figure 2, variously sized increments on the x- and y-axes result in a broken graph. Of particular note was the students' use of the values in the table as the values on the x-axis (1, 15, 45, 60, and 100).

In Figure 3, the student's y-axis is marked by unequal and descending values, resulting in a graph that slopes downward. The student also decided to graph time versus calories rather than calories versus time.

Figures 2 and 3 raised puzzles that fueled lively and thoughtful discussions in class. Together students investigated Figure 2's broken graph and Figure 3's downward slant. By doing so, a teachable moment about the importance of the axes' scales emerged.

Figure 1

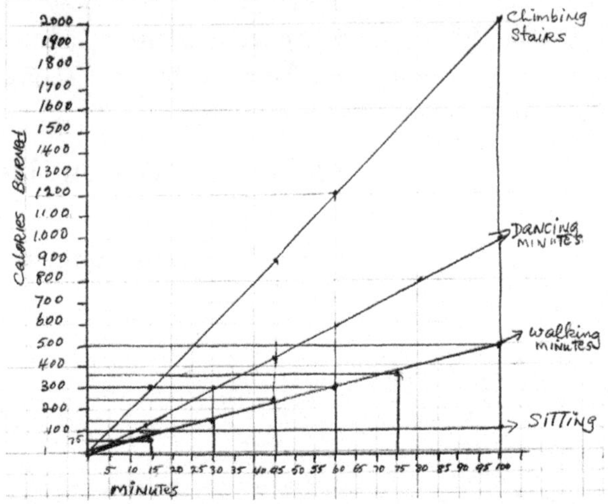

Figure 2

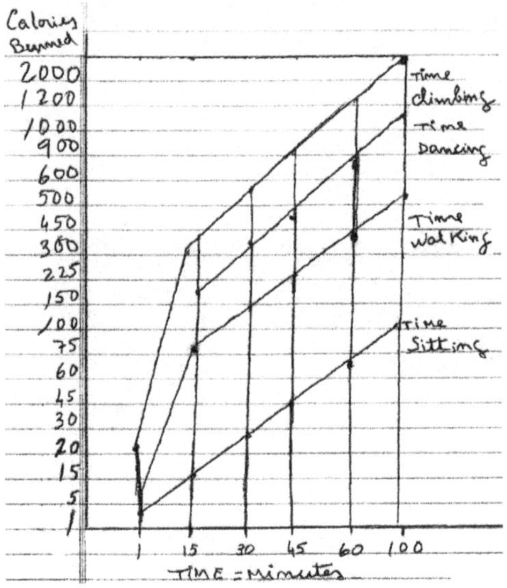

Figure 3

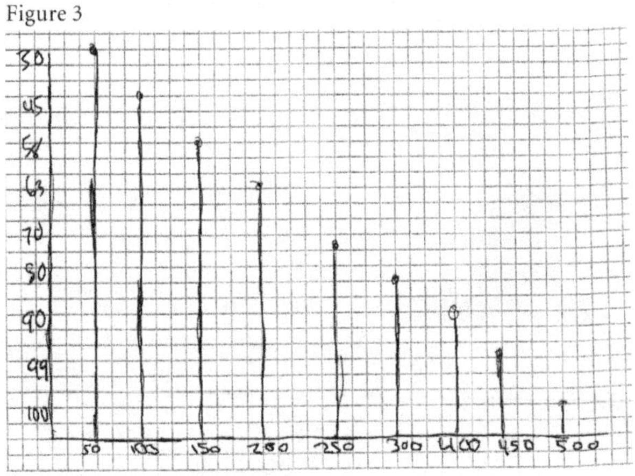

Student work from The Harvard Bridge to Learning and Literacy, Cambridge, MA, and the Borough of Manhattan Community College, NYC

FACILITATING LESSON 5

Body at Work—Pushing It to the Max

What is your average maximum heart rate?

Synopsis

This lesson revisits the topic of graphing from *Lesson 4* and the context of heart rate from *Lesson 3*. Students investigate the connections between a given table, its graph, and its rule, for a situation where one quantity decreases as another increases.

1. Student pairs complete a table of average maximum heart rates for specific ages and write about the patterns they notice (Questions 1–5).
2. The class shares pattern descriptions.
3. Student pairs work on graphs of the data and algebraic rules.
4. The class shares graphs and algebraic rules, continuing to connect the table, graph, verbal description, and algebraic rule.
5. Partners compare this graph, table, and rule with the "Calories Burned Running Up Stairs" graph, table, and rule from *Lessons 3* and *4*.
6. Students make summary statements about what they learned.

Objectives

- Make graphs from tables
- Articulate how graphs, tables, and rules relate to one another
- Compare representations of a situation in which y-values decrease as x-values increase with another in which both values increase together

Materials/Prep

- Calculators
- Colored markers
- Graph paper
- Easel paper
- Rulers

Enlarge a copy or make a presentation slide of the table of average maximum heart rates by age for the whole class to see. Also, set up the board with headings: "Table," "Graph," "Verbal Description," and "Algebraic Rule."

Opening Discussion

Inquire:

Have you ever worked your heart so hard it was pumping wildly, and you thought you might pass out? If you have, you know what it feels like to push your heart to the max, its maximum capacity.

In today's lesson, we will look at average maximum heart rate data that reveal some interesting patterns.

Activity 1: Average Maximum Heart Rates

Refer students to the table of average maximum heart rates by age (*Student Book*, p. 66), and say:

What is the pattern these medical researchers were investigating?

The pattern compares *age* and *maximum heart rate average.* Clarify the term "average maximum heart rate" (AMHR) and highlight that AMHR

- Is the highest number of heartbeats a person can experience without feeling weak or lightheaded;
- Occurs after strenuous exercise (for your age);
- Is a figure determined by testing a lot of people and then averaging their data.

Completing the table

Ask students to begin working in pairs on the first five questions that focus on completing the table.

Bring the class together, collect their predictions, and post all suggestions on a table for the whole class to see. If there are disagreements, encourage discussion.

Age (in years)	Average Maximum Heart Rate (beats/minute)
0	220
10	210
15	205
20	200
25	195
30	190
35	185
40	180
45	175
50	170
55	165
60	160
65	155
70	150
75	145

Direct students' attention to the patterns they saw looking across the rows and down the columns (Questions 4 and 5). Post students' verbal descriptions of these patterns. Those who focus on the patterns in the columns might notice

- Every time the age goes up five years, the AMHR goes down five heartbeats.
- The younger you are, the faster your heart rate.
- Up one year in age, down one beat in heart rate.

Those who focus on the patterns across rows might notice

- The amounts add up to 220.
- Subtract the age from 220 to get the heart rate.

Ask:

Where did you see that pattern?

Continue until every observed pattern has been connected to the table data. Focus attention on the relationship between the two variables, age and AMHR, by asking:

What pattern do you notice when you look across the rows?

That the two amounts add up to 220 may not be an evident pattern. If no one detects that relationship, bring it up after students work on Questions 6–8, where they will make a graph and write a rule. Allow ample time for people to puzzle over the relationship between the variables, using both the graph and the table as tools. Ask pairs who finish early to construct their graphs on the board or easel paper.

Heads up!

Decide as a class which symbols or letters will be designated for each of the variables—i.e., which will be used for age (probably *A*) and which will stand for the average maximum heart rate (probably *H* for heart rate).

Examining the graphs

After students have worked for a while, focus the class's attention on the graphs posted at the front (Question 6). Ask whether anyone has suggestions for improving them and whether anyone had a different-looking graph. This will be a chance to rectify errors such as unclear labeling or uneven increments, which might result in other than a straight line.

Ask:

Looking at the graph, what do you notice?

Where are the table patterns in the graph? For example, where can you see that as the age increases by one year, the heart rate decreases by a year?

Stating the rule

To review Questions 7 and 8, suggest that several students come to the board and post their verbal and algebraic rules for predicting average maximum heart rate for any age. Choose students whose equations or verbal rules differ. The key point to push for is the additive relationship that can be expressed with addition or subtraction:

- $A + H = 220$
- $H = 220 - A$
- $A = 220 - H$

Suggest that students substitute any set of table entries into all equations.

This is an opportunity to demonstrate the relationships among the multiple forms of an equation. Start by asking students to explain each representation in simple terms. Once they have done so, ask:

We see that all three equations work with our table. Why is this so? How are the equations alike, and how are they different?

Point to the connections between $H = 220 - A$ and $A = 220 - H$, where the values subtracted are switched, and $A + H = 220$, where the two variables are added to produce a constant amount.

If everyone was able to make the connection between labeled points on the graph and table entries, press students to connect the graph to the rules. Ask:

Show me where you see the equation in the graph.

Return to the situation of the problem, and ask:

What did the researchers conclude about the relationship between age and average maximum heart rate?

Activity 2: Comparing and Contrasting Situations

The point of this examination is to compare representations of a situation where one variable decreases as the other increases with situations where the variables increase together.

Refer students to *Comparing and Contrasting Situations* (*Student Book*, p. 68). Tell student pairs to refer to the graph "Calories Burned Running Up Stairs" (*Student Book*, p. 53) from *Lessons 3* and *4*. Ask them to list all similarities and differences between that graph and the graph they made today. Then compare the corresponding tables and equations.

Bring the class together to share what they notice. They might include the following:

SIMILARITIES IN THE GRAPHS:	DIFFERENCES IN THE GRAPHS:
Both graphs are straight lines.	The AMHR line slants down; the calories line slants up. The calories graph starts at the origin; the AMHR at (0, 220). The scales are different. The labels are different.
SIMILARITIES IN THE TABLES:	**DIFFERENCES IN THE TABLES:**
Each has two columns.	In "Calories Burned," the numbers in both columns increase, but in "Average Maximum Heart Rates," the second (Out) column amounts decrease.
SIMILARITIES IN THE EQUATIONS:	**DIFFERENCES IN THE EQUATIONS:**
$C = 20m$ $H = 220 - A$ Both have equals signs. Both have two variables (letters).	Addition/subtraction versus multiplication Different numbers and letters
A GENERAL STATEMENT COMPARING THE TWO SITUATIONS:	
As the minutes increase, so do the number of calories. However, as the age increases, the heart rate number decreases.	

Summary Discussion

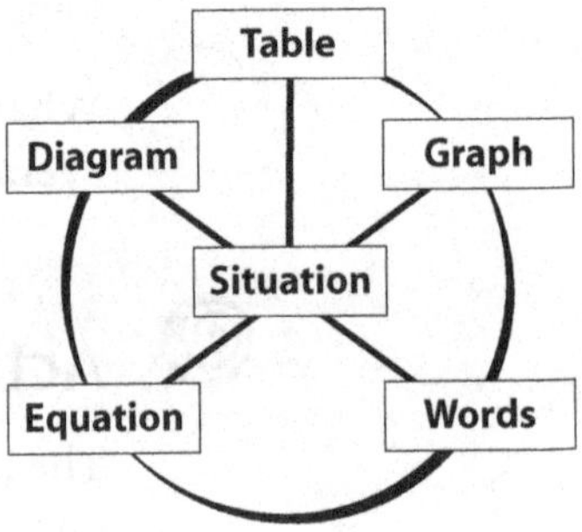

Ask partners to share three important ideas they learned today. After two or three minutes, reconvene the class to share the ideas learned.

Practice

Brick Piles, p. 70

For more practice with a linear relationship, where as one set of values increases, the other decreases.

Symbol Sense Practice

Matching Equations, p. 72

For symbol sense practice with the inverse operations of addition and subtraction in equations with variables.

Extension

Aerobic Target Heart Rate, p. 73

For an opportunity to work with percents.

Test Practice

Test Practice, p. 74

Looking Closely

Observe whether students are able to

Make graphs from tables

This is the second class where coordinate graphing is used. Who has a hard time plotting points, determining increments, or labeling the axes? A ruler may help students plot the points accurately. Suggest that students count up on each axis before they decide on increments.

Remind them that the x-axis is conventionally reserved for independent variables, the variables that cause the change, such as age or time. The y-axis is reserved for the dependent variables, those values that are affected by the independent variable and changed as a result.

Articulate how graphs, tables, and rules relate to one another

What kind of connections are students making? Labeling points on the graph helps them see the connection between the table and the graph more explicitly. Checking rules with the values from labeled points also verifies findings.

Compare representations of a situation in which *y*-values decrease as *x*-values increase with another in which both values increase together

What evidence was there that students saw that the *direction* of a graph's slant told the story that one quantity *increased* as the other *decreased* (as opposed to the graphs in *Lesson 4* where as one quantity increased, the other increased as well)? Again, labeling points provides numerical references to see increase and decrease on the graph.

Are the *y*-values increasing or decreasing, as the age, or In numbers go up? By finding the corresponding pairs of numbers on the appropriate graph, students start to see the numerical relationships in a graph.

WHAT TO LOOK FOR IN *LESSON 5*	WHO STANDS OUT? (LIST STUDENTS' INITIALS)			NOTES FOR NEXT STEPS
	STRONG	ADEQUATE	NEEDS WORK	
Representations Tables • Recognizes vertical and horizontal patterns • Identifies relationship between *x*- and *y*-values • Predicts missing values accurately Graphs • Labels increments consistently and accurately • Labels axes • Explains the cause for the graph's downward slant Rule • Accurately describes relationship between age and average maximum heart rate Equation • Writes an equation that accurately reflects the relationship between age and average maximum heart rate				
Connections • Uses table data to accurately plot graph points • Identifies pattern of decrease in table and graph • Makes comparisons between representations for patterns of increase and patterns of decrease				

Rationale

The introduction of a decreasing pattern situation helps build interpretive capacity. It becomes evident that not all graphs are alike, nor are all equations or tables. The idea that x- and y-values always increase together is challenged as students observe x- and y-values relating to one another differently in the table and on the graph. Students become more aware of the broad range of possible situations that algebra can describe.

Math Background

This lesson takes an intuitive approach, laying a foundation for further formal study of slope and the relationships between variables.

Relationships may be *direct* or *indirect.* On the one hand, if the larger values of variable A are related to larger values of variable B, the relationship is *direct.* For example, as the number of hours worked increases, workers earn more wages. On the other hand, when larger values of variable A are related to smaller values of variable B, the relationship is *inverse* or *indirect.* For example, when the price of beef rises, people buy less of it.

Relationships may be *linear* or *nonlinear.* Linear relationships can be characterized by their slopes and by their positions with respect to the y-axis. In the equation
$y = mx + b$, m indicates the slope and b the y-intercept. In a graph, if m is positive, the line will slant up to the right, a *positive slope.* That is a direct linear relationship. If m is negative, the line will slant down to the right, a *negative slope.* Since it is not assumed in this unit that students have a command of negative numbers, the focus is not on negative slope, even though $H = 220 - A$ can be interpreted as a linear equation in the form of $H = -A + 220$.

Nonlinear relationships do not result in straight lines. Examples of nonlinear relationships are equations such as $xy = k$ and $y = kx^2$, where k is a constant. Both are graphed as curves. A few examples of curvilinear relationships will be touched on in a later lesson as a contrast to linear relationships.

Relationships may be *directly proportional* or *inversely proportional.* Both relationships can be illustrated with the *distance* = *rate* x *time* formula, depending on which variable is held constant. If the rate (speed) is held at 50 mph, then as time increases, so does distance. Moreover, this relationship is linear with a slope of 50 and a y-intercept of 0. However, if the distance is held at 100 miles, as the speed increases, time decreases. The resulting graph is nonlinear and represents an inversely proportional relationship.

In summary, the number of calories (C) is directly proportional to the number of minutes exercised (m). There is a direct linear relationship between calories and minutes. There is also an indirect linear relationship between age and average maximum heart rate. This relationship is not proportional.

Context

The general rule introduced in this lesson (subtract your age from 220 to arrive at the average maximum heart rate) is commonly used by fitness professionals. Its accuracy can vary ± 10 heartbeats. More precise formulas also take weight and gender into account. For example, one formula used is the following:

For males: maximum heart rate = 210 – ½ age – 5% of body weight + 4

For females: maximum heart rate = 210 – ½ age – 1% of body weight + 0

(Source: https://www.cdc.gov/physicalactivity/basics/measuring/heartrate.htm)

Another takes into consideration heart rate at rest to arrive at a target heart rate:

maximum heart rate = 220 – age + resting heart rate

(Source: http://www.intervalsforcardio.com/karvonen-formula.html)

Various training programs use the maximum heart rate and factor in the intensity of the exercise and the heart rate at rest to arrive at particular target heart rate zones. This is known as the Karvonen formula:

target heart rate = % intensity x (age-related maximum heart rate – resting heart rate) + resting heart rate

For the "aerobic" zone, use 70% to 80% intensity.

For the "fat-burning" zone, use 60% to 70% intensity.

(Source: http://www.intervalsforcardio.com/karvonen-formula.html)

Facilitation

Making the Lesson Easier

If students struggle with seeing patterns and making predictions in Questions 1–5 of *Activity 1*, look at the table as a class to determine the pattern and make a prediction. Make a separate table with everyone's age as the In number and average maximum heart rate as the Out number.

Making the Lesson Harder

Introduce the formula that takes into account weight, age, and gender (shown above in *Context*). Ask students to compare the results using this more complex formula to the results using the simpler formula in *Activity 1*.

Introduce the notion of target heart rates for particular exercises. Assign *Extension: Aerobic Target Heart Rate* (*Student Book*, p. 73).

LESSON 5 IN ACTION

Sometimes careful listening and questioning fosters precision and accuracy. One teacher wrote:

Karen struggled to state a rule for finding the average maximum heart rate (H) for a person of any age. She quickly and accurately calculated H for a person of any *particular* age. Her calculations showed that for a 19-year-old person, H would be 201. In a vertical column form, she had subtracted 19 from 220 to solve the problem. In fact, she thought she knew the rule: "Take their age and subtract their heart rate." Karen was ready to move on when the teacher asked her, "Does your rule work with specific figures?" As Karen applied a second set of numbers to her rule, she realized it did not work.

"How did you figure out the average maximum heart rate for someone 19 years old?" the teacher probed.

Karen pointed to her paper: "Like this … 220 – 19 = 201."

"Does what you did match your rule?"

Karen quickly saw it did not. "No, no. I subtracted the age from 220, not the heart rate." A more precise and workable rule was emerging. When Karen substituted other ages in her new rule, she found that it worked. To find the average maximum heart rate for any age, "You subtract the person's age from 220."

Tricia Donovan
Pioneer Valley Adult Education Center, Northampton, MA

Uncovering, articulating, and representing patterns is a multi-step process. Here is how it unfolded in one classroom.

First, the teacher posted a table of average maximum heart rates for a variety of ages. The teacher, Charlie, asked people to predict the average maximum heart rate for a person age 75, a child at birth, and a person their own age.

What patterns or rules do you see? Students' first rules focused on the vertical pattern change in the table: As the age increases, the AMHR decreases. But what is the relationship betweeen any age and the AMHR?

Table of Average Maximum Heart Rates

AGE (in years)	Average Maximum Heart Rate (in Beats per minute)
20	200
25	195
30	190
35	185
40	180
45	175
50	170
55	165
60	160
65	155
70	150

continued on next page

continued from previous page

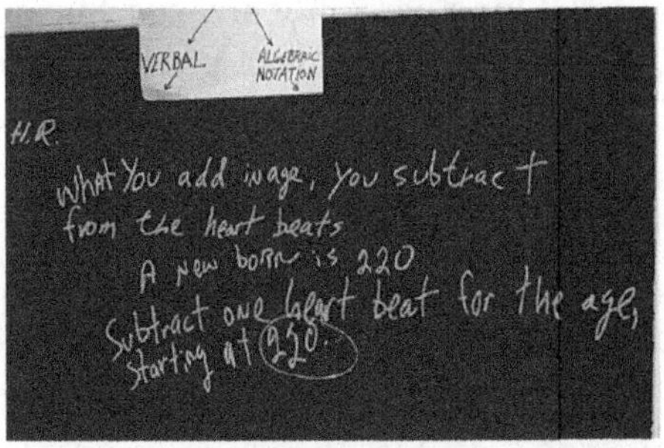

Groups worked to express the relationship between age and the AMHR. They made graphs from the table and looked for the rule in the table and in the graphs.

Success! A student explained with a graph and a rule the relationship that her group arrived at: $220 - x = y$.

BEGIN Managed Programs, New York City

FACILITATING LESSON 6

Circle Patterns

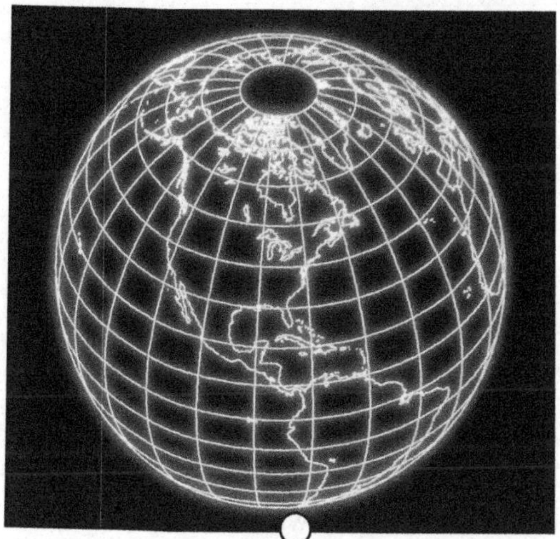

How far is it through the center of the Earth?

Synopsis

Students make connections between algebra and geometry as they gather empirical data and extrapolate approximated formulas involving circumference and diameter. The relationship becomes more precise with the introduction of *pi*.

1. You introduce an investigation about the Earth's size to the class.
2. Student pairs sketch the problem and collect data on circular objects.
3. The class creates a table of everyone's data, discusses patterns, and posts verbal rules for finding unknown diameters and circumferences, before solving the final problem about the Earth.
4. The class discusses *pi* and then practices using 3.14 and *pi* on the calculator to refigure circumference measures.
5. You review the lesson and students enter vocabulary and record what they learned in the *Reflections* section of their books.

Objectives

- Diagram and model physical/quantitative situations
- Describe the approximate relationship between the diameter and circumference of circles
- Apply rules and formulas to solve a problem
- Understand the meaning of *pi*

Materials/Prep

- Calculators
- Circular objects of various diameters from large to small, with a few standard-size objects such as 10-inch pie pans and 9-inch cake pans, enough for everyone to measure two
- Colored markers
- Graph paper
- Easel paper
- Oranges (two) or other round fruit
- Rulers and yardsticks
- String and long pipe cleaners
- Tape

Opening Discussion

Inquire whether anyone has ever done an experiment and what was studied. Explain that scientists often study phenomena that they are *not* able to experience directly, such as conditions on Mars. What they do in such cases is apply what they do know about similar conditions on the Earth, something the class will do today. Ask:

What would you say if someone asked you the distance through the center of the Earth?

Ask students to call out their guesses and record them on the board. Then tell the class that they will engage in scientific and mathematical activity as they work to discover what the Earth's diameter is. They will apply what they already know about algebra to learn more about geometry and the Earth.

Introducing "Circumference" and "Diameter"

Physically demonstrate the **circumference** and **diameter** of a circular object. Have on hand one orange cut in half and one whole fruit. Ask students to imagine that the whole orange represents the planet Earth. Show the orange cut in half through the center.

Say that the line that goes around the Earth's middle is called the **equator**. Point out the equator on a globe, if one is available. Then ask whether anyone knows what the edge of a circle is called in mathematics (**circumference**). Write all terms on the class vocabulary list posted up front.

Place a piece of string or a pipe cleaner across the surface of the halved orange and ask:

Who knows what this line is called in mathematics? (diameter)

The volunteer should show the diameter on the orange by tracing it with a finger. Explain that this information will be useful in the following exercise.

Activity 1: Journey Through the Center of the Earth

Direct attention to *Activity 1: Journey Through the Center of the Earth* (*Student Book*). After the warmup on p. 79, continue to *Investigating Circles* on p. 80. Read the problem together.

Ask pairs of students to work through the problem one step at a time. Remind everyone to begin by sketching the problem. Point out that students should measure carefully in inches (to the nearest ½-inch or ¼-inch, depending on their comfort measuring). Allow 20 minutes for students to work on Questions 1–6.

If some students finish measuring their circles before the rest of the class, ask them to measure one or two more circles.

When everyone has measured at least two circles' diameters and circumferences and has begun collecting data from other students, set up a whole-class table on the board, with one column stating the name of the object, the second the diameter, and the third the circumference. Suggest that those students who are finished post their data.

A sample from one class looked like this:

Object	Diameter	Circumference
Dan's pan	8 1/2"	25 3/4"
Amy's can	3"	8 1/4"
Tricia's cap	3 1/4"	9 3/4"
Patti's pie pan	9"	27 3/4"
Cake pan	9.3"	29"
Lid	5 1/2"	17"
Paper cup	2 1/2 - 2 3/4"	7"
Medium lid	9 1/2"	30"
Other cake pan	7 1/2"	23 1/2"

Now, draw everyone's attention to patterns in the table. Ask:

What pattern do you notice? Where do you see it?

You want students to see that the circumference is always *bigger* than the diameter, and it is *always about three times the diameter*. If no one notices this from the table numbers, have them look at the string lines and ask:

How is the diameter related to the circumference?

Encourage everyone to check data by physically folding the circumference line against the diameter line. It should fold three times with a little left over. Students can also make a physical comparison by taping the diameter string above the circumference string, which will always be about three times as long as the diameter string (shown at right).

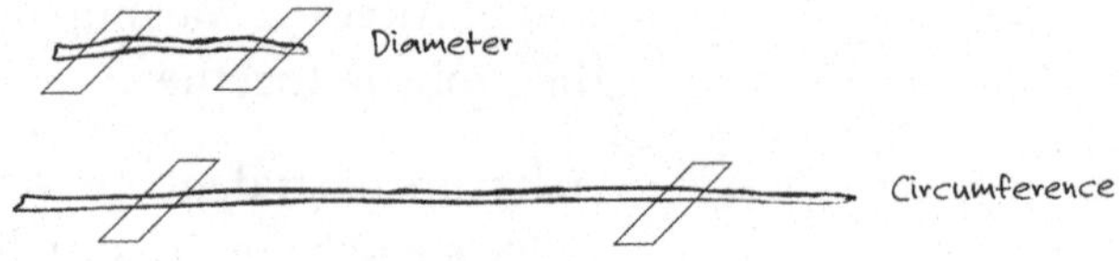

When students begin to see that there are always about three diameters to every circumference length, ask them to confirm this for every entry in the table.

Are there any numbers here that do not fit that pattern at all?

This question may prompt some re-measuring. Allow the class to reach consensus, then ask:

If you know what the diameter measures, how would you make a good estimate for the circumference measure?

If you know what the circumference measures, how would you make a good estimate for the diameter measure?

Post several verbal rules for each situation. Test the rules by measuring the diameter and circumference of the orange. To practice the rule, insert the value "150 feet" in the diameter column, and ask whether anyone knows what the approximate circumference would be and how he or she knows.

To reverse the rule, insert the value "90 feet" in the circumference column, and ask whether anyone knows what the diameter would be approximately and can explain how he or she knows.

Point out that although it is not true that "three diameters equals the circumference" or "$C = 3d$," you could say that the circumference is *about* three diameters or the diameter is *about* one-third of the circumference. These are rules of thumb people use to approximate the circumference or diameter. Estimates like these can be written with two wavy lines.

$$C \approx 3d \qquad \text{or} \qquad d \approx \frac{C}{3}$$

Back to Earth

Now return to the original problem about the Earth. Ask for the length of the Earth's diameter and the ways people figured it out (Questions 7 and 8, *Student Book*, p. 81).

Activity 2: Measure, Calculate, and Compare

Ask whether anyone knows what the number is called that is a little larger than three and that gives a circle's circumference when you multiply it by the diameter's measure. Some students may recall something about *pi* from past math classes.

Write the *pi* symbol on the board and explain that this Greek letter, *pi*, written as π, stands for that number equal to three plus a little.

$C \approx 3d$ is written as $C = \pi d$ to be exact. Sometimes the decimal figure 3.14 is used as a good approximation of *pi*. Point out that *pi* is an interesting number, an infinite nonrepeating decimal figure: 3.141592653589793238462643…

Instruct the class to take diameter measurements of the variously sized circles and to try finding the circumference measurements using the "π" key on the calculator. Answers should be fairly close to the measured circumferences.

Refer students to *Activity 2: Measure, Calculate, and Compare* (*Student Book*, p. 82), where they compare the measured circumference, the calculated circumference using 3.14, and the calculated circumference using the "π" key on the calculator.

Summary Discussion

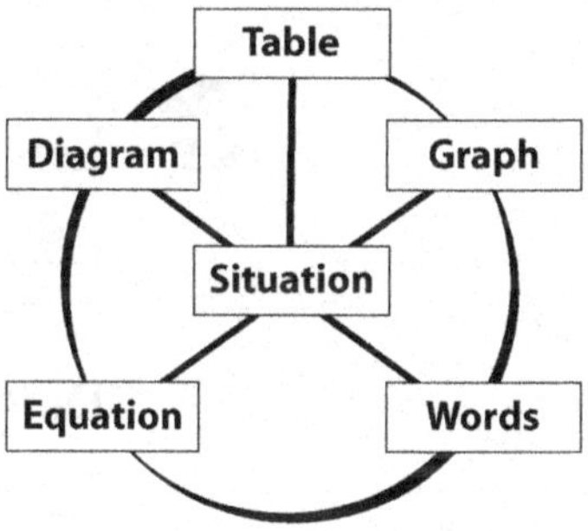

Summarize the lesson by explaining that the class has used information from real objects, diagrams, and tables to figure out the distance of an unseen diameter, just as scientists do. Then ask:

How certain are you that the Earth's diameter is about 7,900 miles long? Explain.

Return to the class's estimates that were posted on the board during the *Opening Discussion*, and take a moment to evaluate them. Close with a review of the problem-solving process. Say:

You were asked a question that is difficult to answer because we cannot go through the center of the Earth. You made some guesses and then collected data about similar situations. Using the circular objects provided, you found patterns and made rules. In geometry these rules are called "formulas." Using formulas helped you find a good approximate answer to the question, "What is the Earth's diameter?" That's the wonder of algebra: no more guesswork!

Remind students to enter the words from this lesson in their *Vocabulary* section (*Student Book*, p. 156), and to record what they have learned today in *Reflections* (*Student Book*, p. 160).

Practice

Graphing Circle Data, p. 83
Asks students to graph circumference versus diameter data from the activity.

How Much More Rubber?, p. 84
Poses a problem to compare the circumferences of two circles, given the diameters.

Symbol Sense Practice

= or ≈ or < or > or ≠ ?, p. 85
Offers practice with relationship notation.

Extension

Investigating Another Pattern in Geometry, p. 86
Asks students to examine the relationship between the number of angles in a polygon and the sum of the angle measures.

Test Practice

Test Practice, p. 88

Looking Closely

Observe whether students are able to

Diagram and model physical/quantitative situations

If a student is unable to sketch the problem, ask this person to describe the situation in words. Encourage the use of a traced circle, cut out and folded in half, to prompt visual awareness. Break the description into parts, illustrating as you go and modeling how you might sketch it.

Describe the relationship between the diameter and circumference of circles

Can students see the rule in the table data? If not, focus their attention on the horizontal entries that most clearly reflect the rule. Ask someone to explain what is happening to the numbers and how that connects to the rules other students stated. Ask students who have trouble generalizing to explain what they see happening to the numbers and how that connects to the rules others stated. It is not necessary at this point for students to have precise notation, though it should be introduced. If someone says $d \times 3 =$ "about the circumference," that is fine.

Apply rules and formulas to solve a problem

Can students solve the Earth problem using the rule they came up with for the circular objects at hand? Have people show the explicit connections between the Earth's measurements and the formula.

Understand the meaning of *pi*.

Can students see by the end of the lesson that *pi* is a constant rather than a variable? Is it clear that the decimal 3.14 is a number greater that 3, but less than 4? The most important idea is that no matter how enormous or infinitesimal the circle, *pi* is always there as the ratio of c/d.

WHAT TO LOOK FOR IN *LESSON 6*	WHO STANDS OUT? (LIST STUDENTS' INITIALS)			NOTES FOR NEXT STEPS
	STRONG	ADEQUATE	NEEDS WORK	
Representations Equations • Writes an equation that accurately reflects the relationship between diameter and circumference • Restates equation using both multiplication and division • Accurately substitutes 3.14 and *pi* into circumference/diameter equations Rules • Accurately describes relationships between diameter and circumference				
Connections • Uses table data to generate and verify rules and equations				
Problem Solving • Accurately applies rules and formulas to solve Earth problem				

Rationale

Connecting mathematical disciplines is an important practice in learning mathematics. In this lesson, students connect geometry and algebra. The focus, however, is on the algebra, finding generalized rules that allow students to predict diameter and circumference for any circle. Moving to large numbers and decimals in this lesson is a reason for supporting work with diagrams.

Math Background

The main point of the lesson is that the ratio of the circumference to the diameter is *always* about 3. This constant ratio often surprises people. Intuitively, some may expect the ratio to increase as the size of the circle increases. Others may think the circumference is double the diameter "since the diameter cuts the circle in half," as one student reasoned.

The exact ratio of the circumference to the diameter is the irrational number *pi*. An irrational number is defined as one that cannot be expressed as the ratio of two natural numbers. In everyday practice, *pi* is approximated to 3.14 or $^{22}/_{7}$, but even carrying out the calculation to the millionth decimal will not exhaust the need for another decimal place.

Although the class discussion is designed to use "about 3" to describe the relationship between circumference and diameter, make sure the point is made that this is an approximation they have discovered, and that scientists make more precise calculations.

Discrepancies in measurement should be addressed, and will vary depending upon the degree of precision. This exercise asks for precision to the nearest half-inch or quarter-inch. If students take an average of three measurements, their recorded measurements will be more precise.

The class will agree upon a number to use for the constant ratio. Most classes will choose 3 as the ratio. Make sure people know this is understood to be "about 3." Other classes may decide upon "a little more than 3" or 3.1 or 3.14, depending on students' math experience. We suggest using "about 3" the first time.

Context

Mathematicians throughout history have studied patterns on land and in the sky to explore numerical relationships. Their findings have given us things like *pi* and the Pythagorean theorem; however, many of these discoveries often overlook where these relationships were first recognized and applied. For example, many mathematicians had a hand in developing the $^{22}/_{7}$ ratio we now know to be constant. Mathematicians in India, Egypt, and Greece all had a part in the development of this very interesting constant ratio that we often simply call *pi*. This ratio is still being explored even today, but with super computers working to find the next digit of *pi*.

Similarly, the Greek mathematician Pythagoras is credited with discovering a unique relationship about the area of a right angled triangle, i.e., the square of the hypotenuse (side opposite the right angle) is equal to the sum of the squares of the other two sides. However, archaeological evidence reveals that this relationship was in use in Babylon nearly 1,000 years before Pythagoras was born.

Si.427, the oldest known example of applied geometry, dating from the Old Babylonian period (1900 to 1600 BCE)

Credit: Daniel.mansfield, CC BY-SA 4.0 <https://creativecommons.org/licenses/by-sa/4.0>, via Wikimedia Commons

Read more about these lesser-known mathematical discoveries and the mathematicians whose names deserve more recognition than previously given. The following resources make for great research and digital literacy explorations with students:

1) *What is Pi?* (https://www.livescience.com/29197-what-is-pi.html)

2) *Australian mathematician reveals oldest applied geometry* (https://news.unsw.edu.au/en/australian-mathematician-reveals-oldest-applied-geometry)

3) *Pythagoras: Everyone knows his famous theorem, but not who discovered it 1000 years before him* (https://www.researchgate.net/publication/247478103_Pythagoras_Everyone_knows_his_famous_theorem_but_not_who_discovered_it_1000_years_before_him)

4) *Timeline of Mathematics* (https://mathigon.org/timeline)

Facilitation

Making the Lesson Easier

Focus on the physical models for diameter and circumference (strings/pipe cleaners) and their relationships to one another, rather than on the table of measurements. Ask each student to share with another student what he or she found.

Keep the numbers simple. Use tape measures to find number measurements of larger circles, and keep to whole number approximations.

Consider putting less emphasis on the numbers in the second activity.

Making the Lesson Harder

After the activity of collecting and comparing measured data, focus the lesson on deriving the diameter with calculations, using 3.14, $^{22}/_{7}$, and *pi* (as used on the calculator). All are in turn only approximations of the nonrepeating, infinite decimal number 3.141592653589793238462643…

Messy Numbers: An Opportunity to See the Fractions

Measurements yield some messy numbers as seen in one class's data below. As a result, teachers probed more deeply about examining the relationships among these fractional measurements. This particular table gave the teacher the opportunity to ask students to explain with both a diagram and a calculator how each measured circumference is a little more than three diameters.

Names	Circumference	Diameter
Wanda	31"	10"
Frank	17"	$5\frac{1}{2}$"
Shari	$23\frac{1}{2}$"	$11\frac{3}{4}$"
Kam	29"	8.3"
Lisa	26.5"	9.3"
Dorothy	$18\frac{1}{4}$"	6"

The Persistence of Perceiving the Diameter as Half of the Circumference

When predicting the relationship between the diameter and circumference, students sometimes predicted the diameter would be about half the circumference. When asked why she thought so, one student said, "Because it cuts the circle in half." Her statement provoked an interesting discussion among the students.

Some students held onto this belief even after taking measurements. Notice Shari's entry in the class table above. The teacher prompted students to double-check each other's measurements.

Harvard Bridge to Learning and Literacy, Cambridge, MA

Flying *C*'s and *d*'s

In this transcript, algebraic notation challenges the class as students struggle to express the rule. First, the teacher refocuses discussion on $C \approx 3d$, rather than on the more difficult division relationship $d \approx C/3$. Later, in the closing, one student continues to have difficulty connecting her confident understanding in words with the symbols.

Teacher: "Looking at these, is there any pattern or relationship that jumps out at anybody?"

Student 3: "Yup, the diameter is three times the circumference."

Teacher: "[Say that] one more time?"

Student 3: "The diameter is three times smaller than the circumference."

Teacher: "Does anybody else see any relationships or patterns that are going on here?" (Silence)

Teacher: "Okay, can we say a sentence turning this around? 'The circumference is …' what if I started with the diameter?"

Student 3: "Three times …"

Teacher: "The circumference is about three times bigger?"

Student 3: "About three times the size of the diameter."

Teacher: "About three times the size of the diameter (Teacher writes this on board). Okay, so how would we say that in a more algebra-type sentence?"

Student 3: "*C* …"

Student 3: "With the sign, you know, it's bigger than…"

Teacher: "So *C* … Actually, come draw it up here…"

Student 3: "Which one is bigger?" (Makes the "less than" sign with her fingers.)

Teacher: "Which one do you think?"

(Student chooses one and writes on board $C > D3$.)

Student 3: "Do I do that?"

Teacher: "Is that three times larger?"

Student 3: "No."

Teacher: "What would three times larger be?"

Student 3: "*C* times … *C* plus *C* plus *C*.

Teacher: "*C* times three?"

Class: "Three … 3*C*… how many times is that?"

continued from previous page

Teacher: "Three times diameter or three times circumference?"

Student 5: "Diameter."

Teacher: "Okay, so how do I show three times diameter? What would I do up here?"

Class: "Times three … three times."

Teacher: "Three times, okay."

Glenn Yarnell,
Holyoke Adult Learning Opportunities (HALO) Center, Holyoke, MA

The Power of Hands-on Activities

For students at all levels of proficiency, the hands-on activity made an impression. In the following classes, students were able to remember the circle relationships, but focused on different aspects of the lesson.

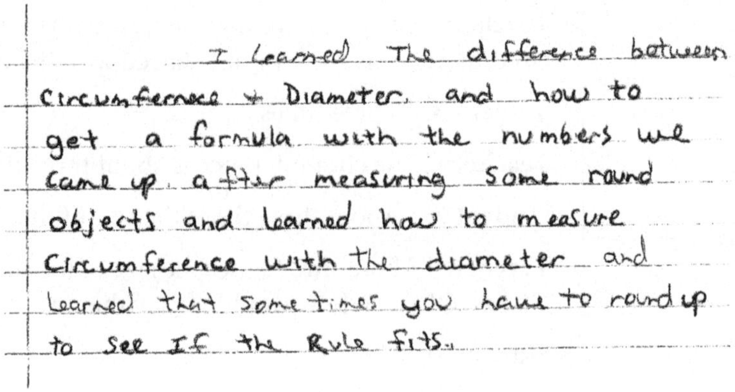

I learned The difference between Circumfernce + Diameter. and how to get a formula with the numbers we came up after measuring some round objects and learned how to measure circumference with the diameter and learned that sometimes you have to round up to see If the Rule fits.

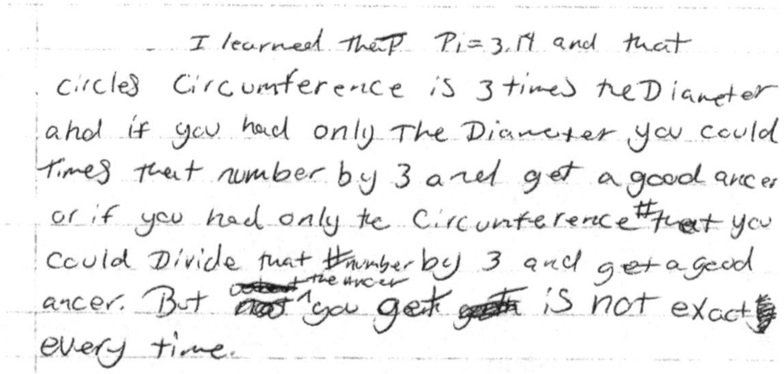

I learned that Pi = 3.14 and that circles Circumference is 3 times te Diameter and if you had only The Diameter you could times that number by 3 and get a good ancer or if you had only te Circumference you could Divide that number by 3 and get a good ancer. But the ancer you get is not exact every time.

Student work from
Mt. Wachusett Community College, Gardner, MA, and
Dimock Community Health Center, Roxbury, MA

Facilitating Interim Assessment: Using the Tools of Algebra

How are things progressing?

Synopsis

This session serves as a review of *Lessons 1–6*. Students work in pairs or individually, depending on how closely you want to observe individual progress. The first four tasks assess how well students are interpreting, producing, and connecting the situation, table, graph, rule in words, and equation. The fifth task assesses students' comfort level with algebraic notation.

Objectives

- Produce other representations of a linear pattern, given a graph, table, or equation
- Interpret linear graphs, tables, and equations, connecting them to the situations
- Interpret and use algebraic symbols

Materials/Prep

- Calculators
- Graph paper
- Rulers

Make one copy of the *Interim Assessment*, pp. 183–189, for each student.

Opening Discussion

Explain that the purpose of today's session is to take stock of what has been covered thus far and to determine students' strengths as well as what areas need more attention before moving on to the next half of the unit.

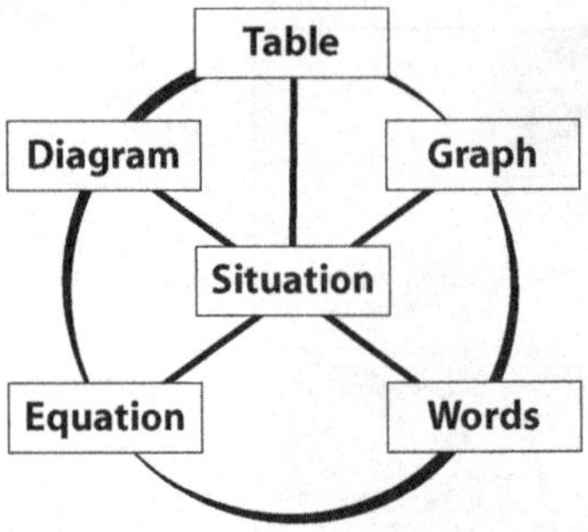

Display the algebra circle diagram (*Student Book*, p. 5), and ask:

How does this diagram show what you have been learning about in algebra?

Listen for responses that mention specific representations and that a main theme is to understand how each representation connects to another. Inform students of the purpose of the assessment:

The assessment tasks will give you the opportunity to show how well you can interpret, produce, and use the different representations to solve problems.

Task 5 will assess everyone's comfort level with understanding and using the symbols of algebra. One option at this point is to suggest that students look back at the *Symbol Sense Practices* in each lesson.

Activity: Assessment Tasks

Distribute *Interim Assessment: Using the Tools of Algebra.* Allow ample time for the assessment, and make sure directions are understood. Ask clarifying questions of students such as

What information are you using to answer the questions?

How are you making predictions for the 10th or 20th entries on your table? How can this help you write a rule?

How can you check your rule to make sure you are right?

Collect all completed work. Plan to give feedback and reflect on the tasks soon after. Use *Looking Closely,* p. 91, and *What to Look For in the Interim Assessment*, p. 93, to guide instructional decisions.

Summary Discussion

Close by reflecting on the assessment tasks. Take a few moments to check in with students.

Which test question did you think was the most difficult? Why?

Which was easiest?

Offer your observations as well.

Compose a list of skills and concepts that students feel they know well and those they think they might need to work on further. Use this data to inform homework assignments and future lessons.

Practice

Any previous assignments that have not been completed.

Looking Closely

Observe whether students are able to

Produce other representations of a linear pattern, given a graph, table, or equation

Creating a table from a graph requires an understanding of the points on a graph—the coordinates of the points become the table data. Ask questions such as "What is this point here?" to see whether students are reading the graph properly and relating the variables.

At first, students will probably develop their generalized rules from the table data rather than the graph. The hotel bill graph data will be easier to state as a rule than the drought table data. It might help to remind students of *Lesson 5, Activity 1: Average Maximum Heart Rates*—did those rules involve multiplication, addition, or subtraction? Direct attention to the table and ask students to consider the relationship between the numbers in the x and y columns.

Similarly, in the vocabulary formula task, the table will be easier to produce than the graph. In fact, the table is a valuable intermediate step and may come in as a useful tool for Tina's walking rate as well.

Interpret linear graphs, tables, and equations, connecting them to the situations

How well are students able to explain the situation that is represented by the tools?

Are students assembling a complete story of the graph? Thoroughness is key here. Students need to take in the information from the title, labels, numbers, and the graph itself. What bits of information are students noticing?

Tasks 3 and 4 focus on equations, a subject that will be explored more fully in the next lesson. However, if students work with the graph or a table of data, they will likely see a solution.

Notice whether students are reading the questions accurately. Do they see the important information? Is notation a stumbling block? Do they read and understand graphs—how line directions will look when extended; how points are generated by substituting values into equations; how graphs reveal patterns?

Interpret and use algebraic symbols

If students have problems with any portion of Task 5, Symbol Sense, suggest they complete, or re-do the relevant *Symbol Sense Practices* in *Lessons 1–6.* You might also ask students who did not have difficulty with the tasks to explain how they did them.

WHAT TO LOOK FOR IN THE *INTERIM ASSESSMENT*	WHO STANDS OUT? (LIST STUDENTS' INITIALS)			NOTES FOR NEXT STEPS
	STRONG	ADEQUATE	NEEDS WORK	
Representations Graphs • Identifies and understands important features in graph • Locates points to establish table data Tables • Constructs a table with appropriate column headings • Identifies relationship between *x*- and *y*-values within the table • Predicts new values correctly Rules • Describes relationship between *x*- and *y*-values in words • Writes relationship between *x*- and *y*-values as algebraic equations • Uses algebraic notation conventions for all four operations when stating equations				
Symbol Sense • Reads and uses algebraic notation conventions appropriately • Recognizes operational inverse relationships (addition/subtraction and multiplication/division) • Applies order of operations accurately				

FACILITATING LESSON 7

What Is the Message?

What are they trying to tell me?

Synopsis

Two key skills are solidified in this lesson: translating between equations and verbal rules; and creating table data, graphs, and contextual situations from equations in two variables.

1. The class reviews equations from situations they have worked with thus far.
2. Student pairs work on translating equations into written rules and vice versa.
3. Student pairs make tables, graphs, and situation descriptions/demonstrations that could correspond to a one-step multiplication or an addition pattern in an equation.
4. The class discusses similarities and differences between students' representations of the equations.
5. The class summarizes the information communicated by an equation.

Objectives

- Translate equations in two variables into words and words into equations
- Generate a corresponding table, graph, diagram, and contextual situation, given an equation in two variables

Photo above: The Enigma Machine, a cipher device developed and used in the early- to mid-20th century to protect commercial, diplomatic, and military communication. Credit: Bundesarchiv, Bild 101I-241-2173-09 / Grupp / CC-BY-SA 3.0, CC BY-SA 3.0 de, https://commons.wikimedia.org/w/index.php?curid=5410514

Materials/Prep

- Calculators
- Colored markers
- Colored tiles, pennies, or other objects with which students can represent patterns
- Graph paper
- Easel paper

Photocopy one copy of *Secret Equation Cards, Blackline Master 4*, for each pair of students. Cut each copy into 10 cards, and put the cards in an envelope marked "Secret Equations."

Photocopy one copy of *Pass the Message Trifolds, Blackline Master 5*, for every student.

Opening Discussion

Reflect back on the many equations the class has worked with in *Lessons 1–6*. Ask for specific examples and record on the board. For example, students might recall:

From *Banquet Tables, Lesson 2, Activity 1*:

$$p = 2t + 2$$

$$t = \frac{p}{2} - 1$$

From *Heart Rates, Lesson 2, Activity 1*:

$$M = 15f$$

From *Calories Burned, Lesson 2, Activity 2*:

$$c = 10m$$

$$c = 5m$$

From *Average Maximum Heart Rates, Lesson 5, Activity 1*:

$$H = 220 - A$$

$$A + H = 220$$

From *Circle Patterns, Lesson 6, Activity 1*:

$$C = \pi d \approx 3d$$

$$d = \frac{C}{\pi} \approx \frac{C}{3}$$

Say:

In past lessons, we started with the situation or the table of values, and ended up with these equations. Today we will start with some equations and work backward to turn each one into a rule in words, a table, a graph, and an imaginary situation.

Activity 1: Pass the Message

Pair up students. Distribute one envelope of "Secret Equations." Review directions for *Pass the Message* (*Student Book*, p. 92).

The pair completes four sheets altogether (two each) of *Pass the Message Trifolds, Blackline Master 5.*

Instruct:

Use complete sentences to write a rule that represents the equation; then fold under the original equation before you hand it to your partner.

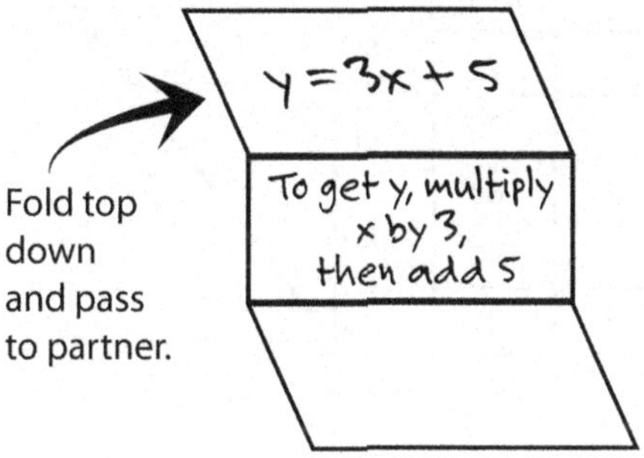

As pairs begin to work, check to see that directions are clear. If many students experience difficulty writing rules, generate and post a list of phrases that come in handy when writing rules, such as

- To find *y*, you …
- *y* equals …
- Multiply *x* by ___, and then add ___ to get *y*.
- Add *x* to ___ …
- Divide *x* by ___, and then …

Encourage pairs who finish early to initiate their own equations. After all the pairs have worked on four equations, gather the class for a discussion. Ask:

What kinds of problems did you encounter when writing the equations from the rules?

Give examples of how you changed a rule to make it clearer.

Heads Up!

To ensure that all student-generated equations are viable, you may want to post originals on the board with all new ones listed underneath. This will trigger a discussion about equivalent statements in which ideas of order and commutativity are likely to surface. For example, does $y = 12x$ mean the same as $12x = y$, or does $y = {}^{x}/_{20}$ mean the same as $y = {}^{20}/_{x}$?

Activity 2: What Is the Message?

Set up the activity by saying:

You know that an equation connects to a rule in words, a table, a graph, and a situation. We will start with an equation.

Post the equation $y = 30x$. Ask one volunteer to start a corresponding table and another to make a matching graph.

x	y
1	30
2	60
3	90
4	120

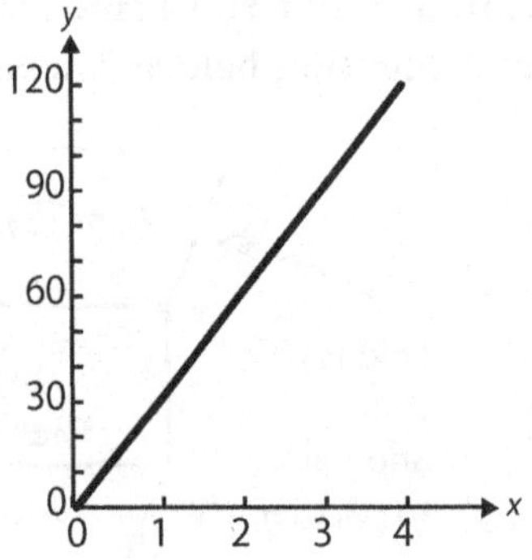

Ask:

What are some real-life situations that could have this pattern?

This question will require creative thinking and careful examination of the table, equation, and graph. Together, generate a list of contexts where the In number (x) is multiplied by 30 to get the Out number (y). Some possibilities are

- If you type 30 words a minute, the number of minutes can be x, and the total number of words (y) will be $30x$.
- If a shirt costs \$30, then two shirts will cost \$30 x 2, and you can get the total cost (y) by multiplying the number of shirts (x) by \$30.
- A driver travels 30 mph. In x hours she travels $30x$ miles.
- With objects, a pattern that grows could look like this:

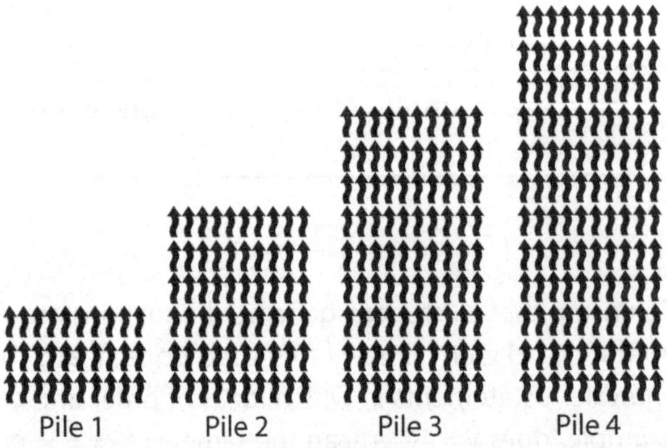

Refer students to *What Is the Message?* (*Student Book*, p. 93). Each pair selects an equation set. Make sure every equation set is chosen. Hand out newsprint and colored markers so pairs can prepare their class presentations for posting.

Remind everyone to work from the multiplication equation first. Review work, prompting for changes if inaccuracies arise, and remind students that several entries in a table and points plotted on a graph are sufficient. Limiting the number of entries will also make transferring the representations to newsprint easy.

Suggest that students work with manipulatives first before diagramming, if diagramming is a challenge for them.

When everyone has posted their work, draw the class together. Center attention on the representations first. Ask:

What similarities do you notice in the graphs? In the tables? In the equations? In the situations?

In what ways are they different?

During the discussion, introduce the term **coefficient** to describe the multiplier and connect its size with steepness of slanting lines.

If no one mentions the following similarities, raise them and post for all to see:

- In the equations, x is multiplied by a number to get y.
- Each successive y term in the table is more than the one before.
- The graphs slant up to the right.
- The greater the coefficient (the multiplier), the steeper the slope.
- The graphs pass through the origin.

Now focus on verifying ways in which the imagined situations and diagrams connect to the representations. Ask:

How did you come up with your situation?

Allow time here for a review of thinking processes. Which representations triggered the situations for students? Was generating a situation harder for some equations than others?

Continue the discussion by selecting one set of representations to discuss. Pose these questions:

Where do you see the multiplication in this diagram?

How does the diagram reflect the situation?

How are the other diagrams and situations similar to or different from this one?

Now focus on similarities in the addition equations. If no one mentions the following similarities, raise them and write them again on the board or newsprint:

- In the equations, the same number is being added to x to get y.
- Each successive y term in the table is more than the one before.
- The graphs slant up to the right, and all the slants are the same.
- None of the graphs start at the origin.

Review the situations and the diagrams as you did for the multiplication equations.

Summary Discussion

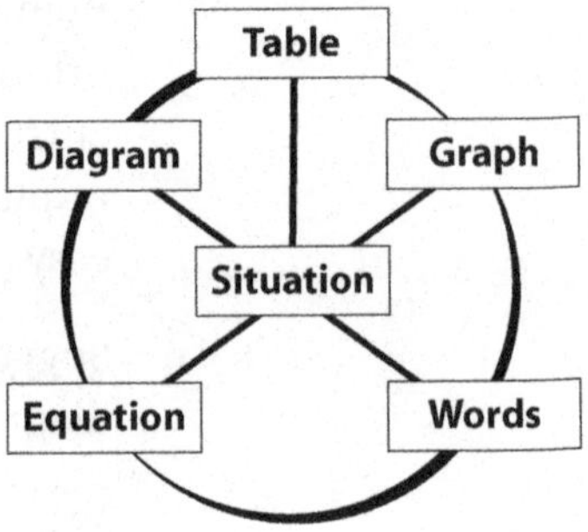

Review the day's mathematics by asking:

How are the situations and representations for multiplication equations different from the situations and representations for addition equations?

Students might conclude:

- In the multiplication situations, the y-values increase faster than in the addition situations.
- In all the addition situations, the y amount increased by 1 each time x increased by 1. In the multiplication situations, the y-value increased by the amount of the coefficient.
- The multiplication graphs pass through the origin; the addition graphs do not.

Invite students to take a few moments to capture their thoughts about today's lesson in their *Reflections* section (*Student Book*, p. 161).

Practice

Geometry Formulas, p. 95
Provides more practice translating equations into words, using common geometric area, perimeter, and volume formulas.

Symbol Sense Practice

Evaluating Geometric Formulas, p. 97
Asks students to substitute and evaluate using common geometric formulas.

Test Practice

Test Practice, p. 98

Looking Closely

Observe whether students are able to

Translate equations in two variables into words and words into equations

Ask students who have trouble writing rules to verbalize what is happening to the quantities in the equation. Suggest they look at the list of phrases generated by the class or, if a list is not generated, at how others wrote rules. "How did this student start his or her rule?" You might also supply a stem, such as, "To find y ...," and let students supply the ending.

Generate a corresponding table, graph, diagram, and contextual situation, given an equation in two variables

How well do students launch the other representations from the equation? How well are they able to:

- Create a table from the equation? Make sure students clearly establish the In (x) and the Out (y) columns, and systematically substitute various values for x, and solve for y, following the relationship stated by the equation.
- Create a graph from the equation? Here the table serves as a valuable intermediate step that students should by now be internalizing. Observe that students are creating axes with clearly marked equal intervals.
- Create a contextual situation, either by diagramming or with verbal descriptions? If students have difficulty with this step, prompt by asking them to label the table with common quantities, such as dollars and pounds, miles and gallons, or hours worked and wages earned.

WHAT TO LOOK FOR IN *LESSON 7*	WHO STANDS OUT? (LIST STUDENTS' INITIALS)			NOTES FOR NEXT STEPS
	STRONG	ADEQUATE	NEEDS WORK	
Representations Equations • Writes rules that accurately represent equations • Writes equations accurately from rules Tables • Generates an accurate table from an equation Graphs • Generates an accurate graph from either an equation or a table Diagrams • Accurately reflects multiplicative or additive relationship described in the equation				
Connections • Clearly connects diagrams to the situation and the relationship described by the equation • Describes situations that clearly reflect the relationship in the equation				

Rationale

Although students have worked repeatedly to generate equations from situations and other representations, they have yet to start from the equation and imagine what it looks like in a graph, table, or diagram, and what kind of situation it might describe. This lesson offers the opportunity to strengthen appreciation for the descriptive features of an equation.

Math Background

The opportunity to talk about equivalent expressions and equations comes up frequently in this lesson. While formal terminology is not emphasized here, algebraic axioms underpin equivalence.

For example,

Reflexivity: If $a = b$, then $b = a$.

$y = 3x \Leftrightarrow 3x = y.$

Multiplicative identity: $a = 1a$.

$y = x \Leftrightarrow y = 1x.$

Multiplicative commutativity: $ab = ba$.

You can equate $3x$ with $x(3)$. However, the convention is to write the numerical coefficient before the letter.

Additive commutativity: $a + b = b + a$.

$12x + 5 \Leftrightarrow 5 + 12x.$

$x + 10 \Leftrightarrow 10 + x.$

Non-commutativity of division: $a/b \neq b/a$.

$y = x/7$ is not the same as $y = 7/x$.

Non-commutativity of subtraction: $a - b \neq b - a$.

$y = x - 5$ is not equivalent to $y = 5 - x$.

Distributivity: $a(b + c) = ab + ac$.

$y = 12(x + 5) \Leftrightarrow y = 12x + 12(5).$

$y = 12(x + 5)$ is not the same as $y = 12x + 5$.

Inverse operations of multiplication and division: If $a = bc$, then $b = {}^{a}/_{c}$ (except when $c = 0$), and $c = {}^{a}/_{b}$ (except when $b = 0$).

$y = 5x$, $x = {}^{y}/_{5}$, and $5 = {}^{y}/_{x}$ (except where $x = 0$) are equivalent equations.

Inverse operations of addition and subtraction: If $a = b + c$, then $b = a - c$ and $c = a - b$.

$y = x - 5$, $y + 5 = x$, and $5 + y = x$, are equivalent equations.

Facilitation

Students usually find it easier to generate multiplication situations than addition situations. Looking at the graph can remind students that in the addition situations, one starts with a given amount and it increases from there.

When generating multiplication situations (such as $y = 7x$) in *Activity 2: What Is the Message?*, ask: "What comes in sevens?"

When generating additive situations (such as $y = x + 7$), ask: "What is a situation where one value is always 7 more than the other?"

Making the Lesson Easier

When translating equations, post a list of sentence stems that can help students start their rules. Do one or two as a class before pairs start to work.

Work only with multiplication equations ($y = 12x$; $y = 7x$; and $y = 40x$) when doing *What Is the Message?*

Making the Lesson Harder

Use more difficult numbers and equations.

Compare multiplication and addition equations together.

Ask why all equations involving addition start on the y-axis. Why is the slant of the line positive (goes up to the right) even with division?

Good modeling can help students gain independence and skill.

One teacher wrote:

When Felipe gave Mike his equation folded over, Mike kept flipping it back and forth to see if Felipe got it right. Then Mike copied the original equation because he agreed with it. I sat with him when Felipe gave him the next one, and I held it down. I asked him to read it as "*y* equals 2 times *x* plus 25." I said, "Okay, now write '*y*,' ... now 'equals,' ... now '2 times *x*,' ... now 'plus 25.'" I showed him he was correct, and he said, "But you did it!" Eventually, however, Felipe profited from the modeling and translated rules into equations correctly—on his own!

Barbara Goodridge of the Lowell Adult Education Center typed up her students' work so everyone could review the many ways of saying the same thing. This provided a great opportunity to review conventions; to note misconceptions (shaded) that may indicate lack of precision with language or lack of awareness about operations; or to compare ways of stating relationships whether in words or in symbolic language. Students also reviewed spelling. The words and equations are as students wrote them.

Equation Translations

Top Equation	Words	Bottom Equation
$y = x - 12$	*y* equals *x* minus twelve	$y = x - 12$
	x subtract 12 to get *y*	$x - 12 = y$
	x minus 12 equals *y*	$x - 12 = y$
	y equals *x* minus twelve	$y = x - 12$
$y = x + 20$	To get *y*, add 20	$y = x + 20$

continued on next page

continued from previous page

$y = 20 - x$	To get y, subtrek twety frum x	$y = 20 - x$
	To get y, put 20 subtract x	$20 - x = y$
	y is equal 20 less x	$y = x - 20$
	y equal twenty subtact x	$y = 20 - x$
$y = {}^{x}/_{7}$	To get y, dived seven into x	$y = {}^{x}/_{7}$
	y equals x divided by seven	$x \div 7 = y$
$y = 12x + 25$	To get y put twelve and multiply by x and add 25	$y = 12x + 25$
	12 multiplied by x plus 25 equals y	$y = 12x + 25$
	Twelve times x plus twenty-five equals y	$12 \text{ x } x + 25 = y$
$y = {}^{x}/_{20}$	x divided by 20 equals y	$x = y\ 20$
	To get y you must divide 20 into x	
	x divided by twenty equals	$x = 20$
	y multiply x $20 = x$	
$y = 12x$	y equals twelve times x	$y = 12 \text{ x } x$
	Twelve by x is equal to y	$12x = y$
	To get y multiply x by 12	$12x = y$
	y equals x multiplied by 12	$y = x(12)$ $y = 12x$
	y equals x multiplied by 12	$y = x \text{ x } 12$
$y = 7x$	To get y, multiply x by 7	$y = x \text{ x } 7$
	y equals seven x	$y = 7x$
	7 times x equals y	$7x = y$

continued on next page

continued from previous page

$y = x$ $(y = 1x)$	One times x equals y y equals x	$1 \times x = y = x$
	x equally y	$x = y$

Barbara writes: "A few students breezed through the equations—so I created three more challenging ones."

$x^2 + 3 = y$	x to the second power plus 3 equals y	$x^2 + 3 = y$
$x^3 + 6 = y$	y equals x to the third power plus 6	$y = x^3 + 6$
$y = 2(x + 4)$	x plus 4 times 2 equals y	$2(x + 4) = y$

Focusing on equations and related representations opens the door to students' reasoning.

Some basic operational misconceptions surfaced. The teacher noticed students accepting $y = 20 - x$ and $y = x - 20$ as equivalent statements. When asked how the translated equation and the original could be considered the same, students explained that, "Well, $x + 20$ is the same as $20 + x$, right?" After working with a few substituted values, the students quickly saw their error.

When making a table of values, one student put all "1's" in the x column. This student had not participated in earlier work on tables. She told the teacher, "I always thought that x was one."

When reviewing the graph lines, one student described the multiplication line as "narrower." According to her teacher, this was "her way of describing the steep rise as opposed to the gradual rise of the addition line." Clearly, this student was focused on the relationship of the graph to the y-axis. Steepness to her meant a narrower space between the two—the axis and the graph.

Barbara Goodridge
Lowell Adult Education Center, Lowell, MA

FACILITATING LESSON 8

Job Offers

How can I compare these job offers?

Synopsis

The usefulness of algebraic tools for solving certain problems becomes evident when two linear patterns with different starting points and intersecting lines are compared. This is the first of several lessons that highlight unique graph features, such as the y-intercept, the point of intersection, the distance between lines, and the pattern of change.

1. After reading aloud a job-offer scenario, students pair up to discuss the problem. Pairs offer initial guesses and then solve the problem and share their conclusions with the class.
2. Student pairs decide on a method for solving another question—whether there ever was a point in time when earnings would be the same for both jobs.
3. The whole class makes connections between the situation and their tables, graphs, rules, and equations. You direct attention to the y-intercept, point of intersection, and rate of change.
4. Students reflect on the ways tables and graphs are useful for solving problems and understanding situations.

Objectives

- Compare patterns in tables, equations, and graphs in order to make decisions
- Connect the features of a graph with the features of an equation, particularly the point of intersection, the y-intercept, the distance between the lines, and the rate at which change occurs

Materials/Prep

- Calculators
- Colored markers
- Graph paper
- Easel paper
- Play money for demonstrations (optional)

Prepare a whole-class graph on easel paper or as a projected presentation slide, showing the QuinStar pay-over-time pattern *without* the sign-on bonus and the LaserLink pay-over-time situation with the sign-on bonus.

Heads Up!

Set up the board with the same headings as in previous lessons: "table," "graph," "verbal rules," and "algebraic rules."

Opening Discussion

Open by asking:

If you had two job offers to choose from, what factors would affect your decision?

Talk about the strong role mathematics plays in the decision-making process. When comparing cases, you must figure out the patterns for each situation under consideration and the ways in which patterns interact and relate to each other. The class will do this in the context of examining two job offers.

Activity: Job Offers

Students work with partners. Match skilled and unskilled graph-makers.

Ask a volunteer to read *Job Offers* (*Student Book*, p. 102). Have each pair briefly discuss whether they agree with Armand or Cheri. Then ask the class:

Do you agree with Armand or Cheri? Why?

Post everyone's votes and briefly discuss reasons. Then pairs of students work on the problem for five minutes before you ask:

How did you decide who was right?

Post one or two of the processes shared. These will most likely describe what was multiplied and added. Save the voting results for later reference.

Extend the problem situation, saying:

Armand would start out making more at QuinStar, but by the end of the year, he would have made more at LaserLink. Would there ever be a point in time when the accumulated earnings from each job would be the same?

Accompany your statements with the diagrams below:

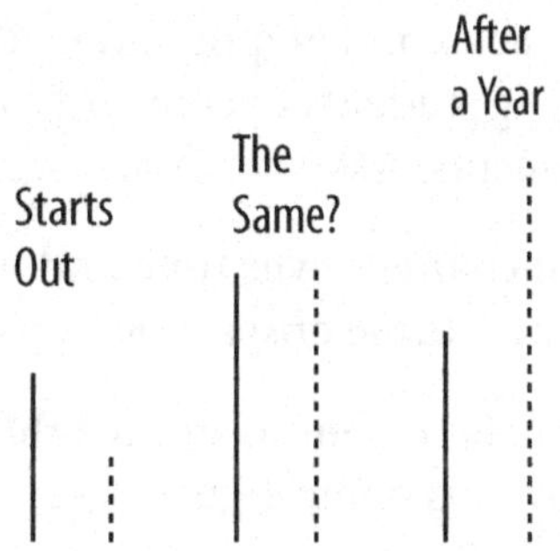

Allow time for discussion. Encourage students to speculate about when this "equal pay" situation might occur, and then tell them that the tools of algebra can help figure out this type of problem.

Would They Ever Be the Same?

Prominently post the question: Would there ever be a point in time when the accumulated earnings from each job would be the same? Ask students to take a few moments to think about which algebra tools might help them see the patterns in this problem. Inquire whether students are thinking about using graphs or tables and how they think each of those tools might help.

Student pairs then work through the questions, *Would They Ever Be the Same?* (*Student Book*, p. 103). Although pairs may choose to start with either a table or graph, they are expected to complete both in order to fully participate in the debriefing.

Heads Up!

Be sure everyone understands that the entry for each week is not the amount earned during that week, but rather, the *accumulated* amount earned up to that time. This amount corresponds to the entry on a pay stub that has the year-to-date (YTD) earnings.

Emphasize that students do not need an entry for every week. As they complete their tables and graphs, ask everyone to describe three things they noticed. Suggest they write about

- What stays the same over time;
- What changes over time;
- Anything that stands out for them.

It is important that pairs finish graphs and tables before you start the discussion. Make a few comments on the problem-solving processes you observed as pairs worked:

I noticed that some people started with a table because they… I noticed that others started with a graph, and they said …

Then choose several accurate tables and graphs for students to post, along with their observations. Ask whether students see these observations in their own tables and graphs and where they see them. Pose the following questions, inviting students to come up front and highlight their tables or graphs to support statements. Ask:

What changes over time and what stays the same over time in this table? Connect these observations to the situation.

Where is the bonus in the table? (Write out each QuinStar entry as a composite of pay plus bonus figures to emphasize the constant, if no one did this.)

What changes over time and what stays the same over time in this graph? Connect these observations to the situation.

Where is week zero for each job?

Where is the bonus in the graph? Why is it here? (0, 2000)

Explain that when the graph starts at a point on the y-axis other than the origin (0, 0), the starting point is called the **y-intercept**. Add that term to the class vocabulary list. Hold off on a discussion of the point of intersection. Instead, focus on the rules:

What is the equation for the LaserLink job pattern?

What is the equation for the QuinStar pattern?

Invite a few pairs to post rules and equations. Discuss where the pay rates and the bonus are in the equation as you connect verbal rules and equations with the solutions written earlier. Say, for example, "So that is why you multiplied 52 weeks by $200." Discuss questions posed by the class until everyone reaches agreement on the equations.

Point to the graphs' point of intersection (at 40 weeks):

I see the bonus here (0, 2000) **and the pay rates here** (keeps going up by 200 or 150)**, but what is going on here?**

Students will most like say the two lines cross at that point, and some should see it as the point where both jobs yield the same income. Volunteers label the point for both graphs. Students then return to the tables they made and connect the point of intersection on the graph to the entries on their tables. Students should reach the understanding that this is the point at which Armand would have earned the same pay at both jobs.

Ask the students:

What does a point of intersection tell you?

Add the phrase **point of intersection** to the class vocabulary list.

Then add that there is something else important and visible in this graph. Ask:

At which job do earnings accumulate or grow faster? How do you know? In the tables? In the graph? In the equations?

The discussion should raise issues such as the differing slants of the two lines, and the connection to the coefficients in the equations. Do not labor over this discussion. It is enough for students to understand that the higher the coefficient (multiplier), the steeper the graph, and the faster the change over time.

Focus next on the space between the two lines.

What information does the space between the two lines tell you?

The varying distance between the lines tells the story over the year: QuinStar would start out paying more than LaserLink, but gradually LaserLink would close in and catch up by the 40th week. From then on, LaserLink would be more and more lucrative. Ask students to predict what the space between the lines would look like by the end of the second year.

Close the discussion by asking:

What would you expect to happen if we substituted the number "40" for the weeks in each of our equations? Why?

Invite the class to substitute the number "40" for the number of weeks to see how it correlates to the graph and table.

Summary Discussion

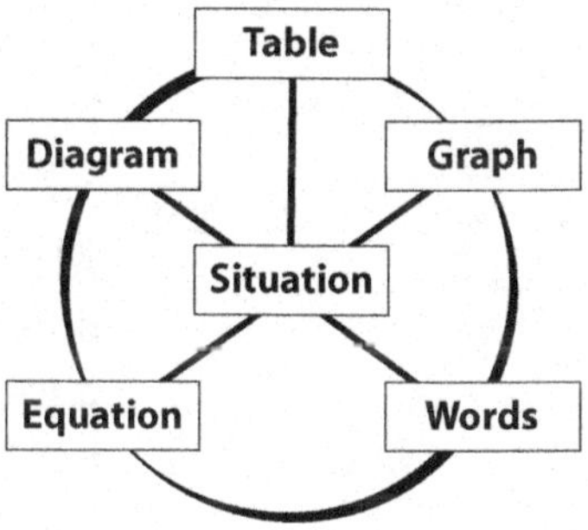

State:

Today you saw that algebraic tools can make solving problems easier.

Ask:

How did the graph, table, or equations help you solve the mathematical dilemma about a time when both jobs would have paid the same?

Invite students to reflect on their own experience and to share one idea they are taking with them from this lesson. Then students enter words from this lesson in their *Vocabulary* section (*Student Book*, p. 156) and write in their *Reflections* (*Student Book*, p. 161).

Practice

The Race, p. 104
Offers a new context, but the same mathematics as *Job Offers.*

Armand's Weeks, Not Pay, p. 105
Turns the multiplicative situation into a division situation.

Symbol Sense Practice

Solving One-Step Equations, p. 107
For practice solving equations in the form of $ax = b$, $x + a = b$, $x - a = b$, and $x/a = b$.

Extension

What If? p. 109
Offers several different scenarios related to *Job Offers* and asks students how the graphs would look for each.

Test Practice

Test Practice, p. 110

Looking Closely

Observe whether students are able to

Compare patterns in tables, equations, and graphs in order to make decisions

Are students making the connection with decision-making? Be sure to ask them to consider who is really right, Cheri or Armand.

Are students having difficulty with the notion of "accumulated" income? When they make tables, do they incorrectly list the weeks and repeat the pay amount for each week? If so, address this problem.

Also, ask students to compare the entries in the table: What is happening to the differences between potential salaries earned at QuinStar and Laserlink? How does this make Armand's decision trickier? What if he were planning to work for less time? More time?

Connect the features of a graph with the features of an equation, particularly the point of intersection, the *y*-intercept, the distance between the lines, and the rate at which change occurs

Are students' graphs correct? Look for problems they are having with the sign-on bonus and where to put it in the table. It should be listed as the "Week zero" entry, as Armand would not have worked at QuinStar yet. Emphasize that the sign-on bonus would be received *before* the first paycheck was issued, so no time would have passed on the weeks-worked axis. Therefore, the graph would start on the y-axis at the point (0, 2000).

Students may see in the tables the time when both job offers would result in the same accumulated pay, but connecting that with the point of intersection in the graph is a real moment of revelation for most. They can then see the graph as a vehicle for telling the details of a situation. Highlighting the entries on their tables and the labeled points cements the visual connection.

Inserting steps—over and up from point to point—on the graphs may help students see the effect of the multiplier (or x-coefficient or rate of change). Ask by how much the graph is going up at each point. It always goes up by the same amount. If students do not say the LaserLink line is going up faster than the

QuinStar line, ask them to return to the table where the salary amounts, not including the sign-on bonus, are listed for both companies. Tell them to consider with which company the money earned increases faster and relate this to the graph, again inserting steps—over one, up 200; or over one, up 150. Label those "up" amounts and discuss how the larger the increase, the steeper the line and, conversely, the steeper the line, the faster the increase.

WHAT TO LOOK FOR IN *LESSON 8*	WHO STANDS OUT? (LIST STUDENTS' INITIALS)			NOTES FOR NEXT STEPS
	STRONG	ADEQUATE	NEEDS WORK	
REPRESENTATIONS Equations • Writes accurate equations for the one-step and two-step situations Tables • Creates an extensive table that communicates information well. Graphs • Obtains information from the *y*-intercept • Obtains information from the intersection of two lines and the space between the lines • Obtains information from the slant of the line				
CONNECTIONS • Explains where the *y*-intercept shows up in the table and equation • Explains where the intersection shows up in the table and equation • Connects the slant of the line to the table data **PROBLEM SOLVING** • Answers the question: Which company would pay more at the end of a year? • Uses change over time to solve the problem				

Rationale

Algebra is used in business, engineering, and science professions to make decisions. Sometimes single situations are studied, but often multiple phenomena are compared. In the latter cases, the focus changes for graphs and tables. Therefore, it is important for students to extend their observational repertoires to include such events as graph lines intersecting and graph lines starting at locations other than (0, 0).

Math Background

This lesson focuses on features of a linear function in an informal, intuitive way. Attention is brought to bear on slope, y-intercept, and solutions to simultaneous equations. These features help tell the story of the situation and are clearly visible in equations through the variation in x-coefficients, which show the speed at which change occurs in graphs.

A more formal study of algebra will connect all cases of the general point-slope form of the linear equation $y = mx + b$ to the graph of a line.

Context

Many people will not have had the opportunity of being offered two jobs at the same time, but they may have had to choose between two long-distance phone carriers, cell-phone programs, or childcare services. The idea of comparison remains the same.

Facilitation

Making Tables

Students approach making a table for Armand's accumulated pay in different ways. Some copy the style used in previous lessons; others make their own variations. It is not important what the table looks like, as long as it organizes the information in an accurate way. Also, some students will want to make 52 entries, whereas others will use proportional relationships or other methods to skip entries as they search for the answer. It is your decision whether to make this efficient model a subject of conversation. The number of entries is irrelevant; however, making an entry for week 40 *is* important.

Making the Lesson Easier

Start with a physical demonstration using play money.

Develop the *Opening Discussion* table and graph as a whole class. Continue to work together as a whole class on the *Activity* if necessary.

Making the Lesson Harder

Challenge students to rewrite the equations for situations where the amount of pay is known and the task is to find the number of weeks. (See *Practice: Armand's Weeks, Not Pay*, p. 105.)

Ask students to consider the graphs, given changes in the situations. (See *Extension: What If?*, p. 109.)

Set up the equations for finding LaserLink and QuinStar totals that equal one another, and discuss possible methods to solve for x in the equation $2{,}000 + 150x = 200x$.

The ways students set up their tables showed various efforts at efficiency. We saw students set up every single entry and set up entries by two-week and five-week intervals, or increasingly larger intervals, once they got the gist (Figure 1). Three- and four-column tables were common.

Figure 1

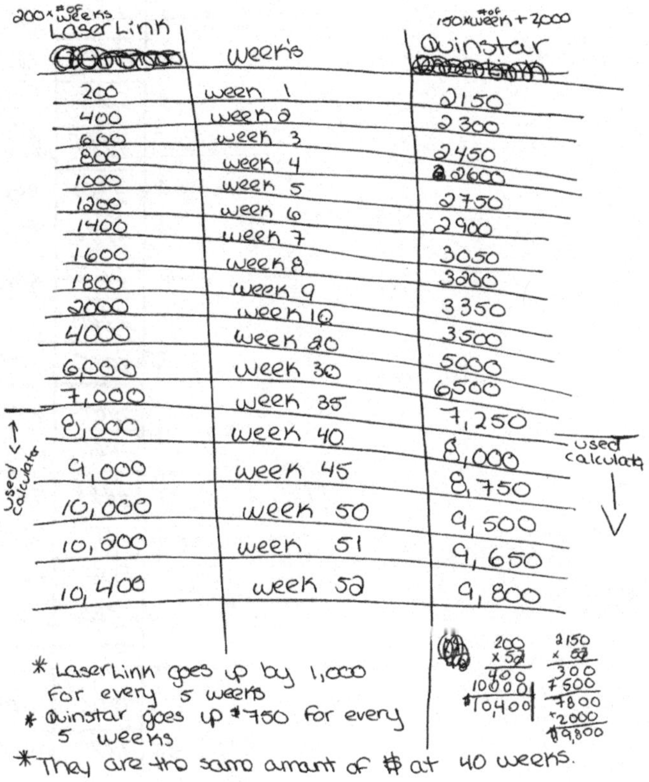

200 × # of weeks Laser Link	weeks	150 × # of week + 2000 Quinstar
200	week 1	2150
400	week 2	2300
600	week 3	2450
800	week 4	2600
1000	week 5	2750
1200	week 6	2900
1400	week 7	3050
1600	week 8	3200
1800	week 9	3350
2000	week 10	3500
4000	week 20	5000
6,000	week 30	6,500
7,000	week 35	7,250
8,000	week 40	8,000
9,000	week 45	8,750
10,000	week 50	9,500
10,200	week 51	9,650
10,400	week 52	9,800

used calculator

200 × 52 = 400 + 10000 = $10,400

2150 × 52 = 300 + 7500 = 7800 + 2000 = $9,800

* LaserLink goes up by 1,000 for every 5 weeks
* Quinstar goes up $750 for every 5 weeks

*They are the same amount of $ at 40 weeks.

One student spontaneously made a table that showed the difference between the two jobs each week. Other students admired his work and asked for a copy. The teacher made copies for all and then asked, "So where can we see these differences in the graph?" (Figure 2).

Figure 2

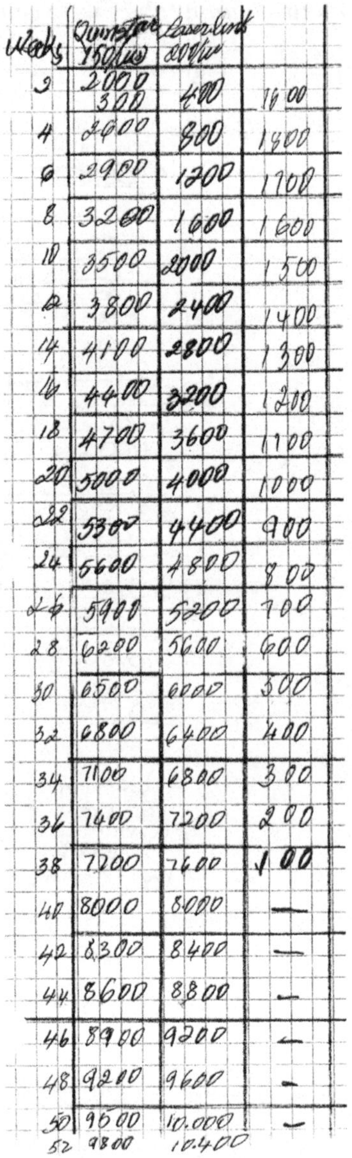

Weeks	Quikstar 150/wk	Laserlink 200/wk	
2	2000 300	400	1600
4	2600	800	1800
6	2900	1200	1700
8	3200	1600	1600
10	3500	2000	1500
12	3800	2400	1400
14	4100	2800	1300
16	4400	3200	1200
18	4700	3600	1100
20	5000	4000	1000
22	5300	4400	900
24	5600	4800	800
26	5900	5200	700
28	6200	5600	600
30	6500	6000	500
32	6800	6400	400
34	7100	6800	300
36	7400	7200	200
38	7700	7600	100
40	8000	8000	—
42	8300	8400	—
44	8600	8800	—
46	8900	9200	—
48	9200	9600	—
50	9500	10.000	—
52	9800	10.400	

Students became facile at coordinating their representations (Figure 3), and evidenced an increasing ability to describe and analyze the representations (Figures 4 and 5).

Figure 3

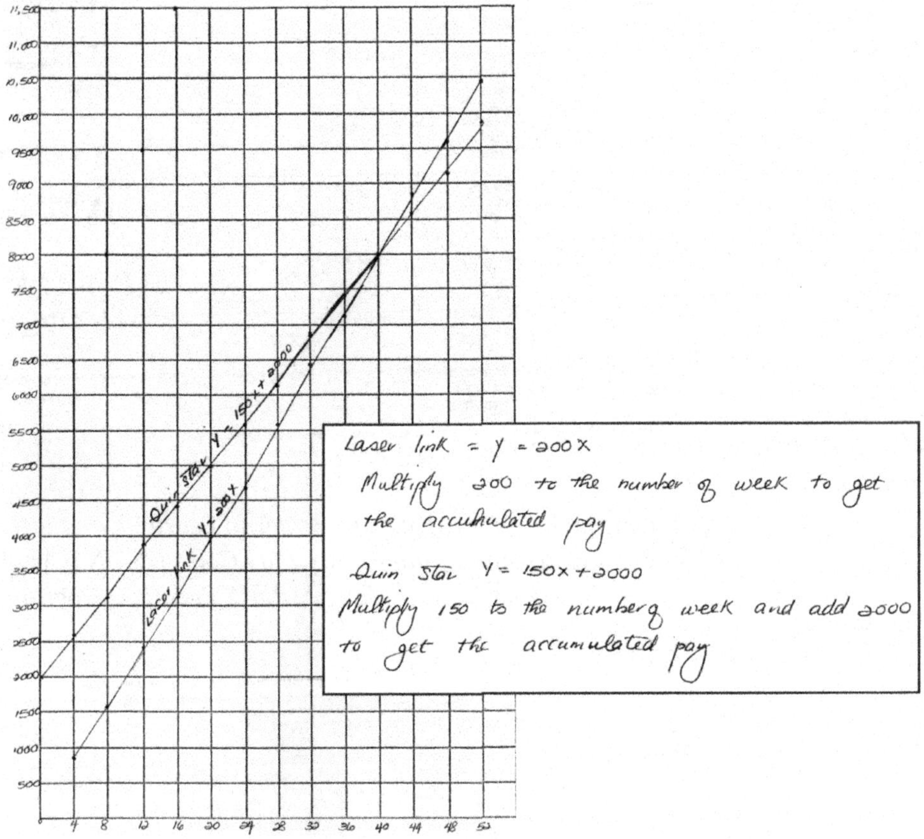

Figure 4

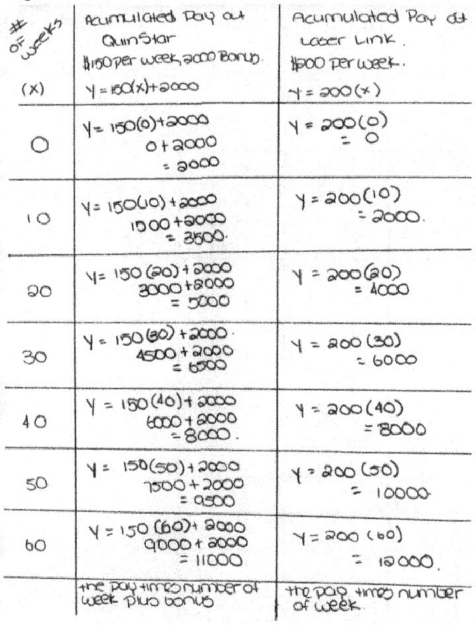

# of weeks (X)	Acumulated Pay at QuinStar $150 per week, 2000 Bonus. Y = 150(x)+2000	Acumulated Pay at Laser Link. $200 per week. Y = 200(x)
0	Y = 150(0)+2000 0+2000 = 2000	Y = 200(0) = 0
10	Y = 150(10)+2000 1500+2000 = 3500.	Y = 200(10) = 2000.
20	Y = 150(20)+2000 3000+2000 = 5000	Y = 200(20) = 4000
30	Y = 150(30)+2000. 4500+2000 = 6500	Y = 200(30) = 6000
40	Y = 150(40)+2000 6000+2000 = 8000.	Y = 200(40) = 8000
50	Y = 150(50)+2000 7500+2000 = 9500	Y = 200(50) = 10000.
60	Y = 150(60)+2000 9000+2000 = 11000	Y = 200(60) = 12000.
	the pay times number of week plus bonus	the pay times number of week.

Figure 5

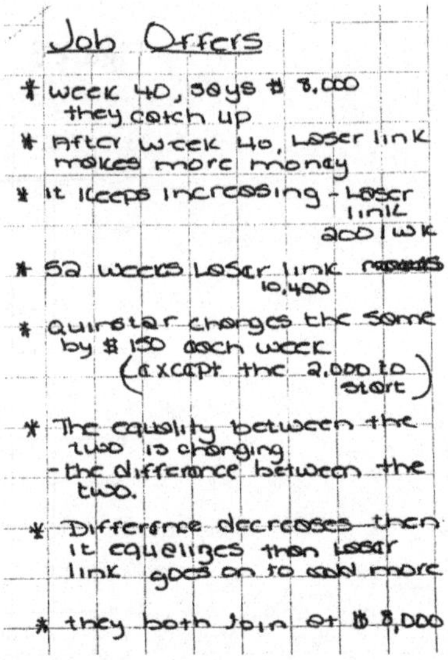

Student work from the Harvard Bridge to Learning and Literacy, Cambridge, MA; Borough of Manhattan Community College/CUNY Adult Education Program; and Pioneer Valley Adult Education Program, Northhampton, MA.

FACILITATING LESSON 9

Phone Plans

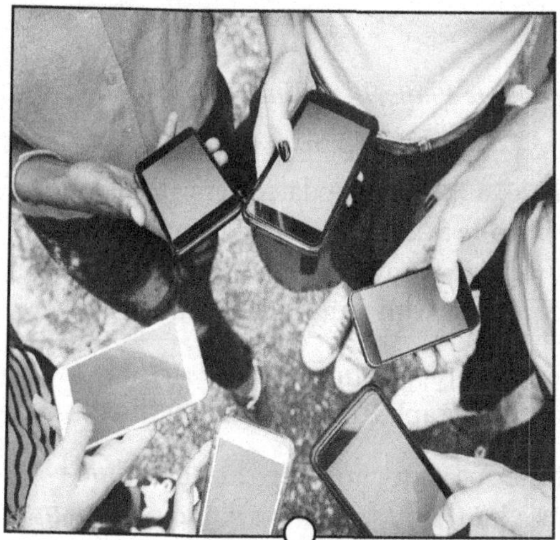

How do I know which plan is best for me?

Synopsis

Students piece together information about four long-distance phone plans and decide which plan best suits three customers with different needs. Previously, situations revealed patterns with a constant increase or decrease; here students also consider a situation where there is no change.

1. Pairs match ads, tables, graphs, and equations for four phone plans.
2. The class comes together to discuss graph and equation features that aid in the matching process.
3. Pairs consider three consumer scenarios to decide which plan would be best and which would be worst for the customer, and justify their reasoning.
5. Everyone reflects on the representations, and the class discusses which tool they would use if faced with a similar situation.

Objectives

- Match graphs, tables, equations, and verbal rules by identifying the related features in each representation
- Connect the flatness of a horizontal line on a graph to a situation in which there is no change over time
- Use information from tables, graphs, rules, and equations to support consumer decisions

Materials/Prep

- Calculators
- Easel paper
- Rulers
- Tape or glue
- *Phone Plans, Blackline Master 6* **or** *Phone Plans, Blackline Master 6a*

There are two versions of the Blackline Masters for Phone Plans. Teachers may choose to use the original (*Blackline Master 6*) and/or the updated version (*Blackline Master 6a*). **Note: the *BLM6a* pages also include Student Book pages and the Answer Key used with this activity).** Both options provide students with similar exploration with ads, tables, graphs, and equations with four phone plans.

Make an enlarged version of each phone-plan graph (*Student Book, Activity 1: Phone Plans*), on easel paper, or reproduce the graphs on presentation slides.

Photocopy the *Phone Plans* Blackline Master of your choice (p. 1 of *Blackline Master 6a*), one for every pair of students. Cut the 12 pieces apart and put them in an envelope. (Students could do the cutting as well, but the scrambled pieces with unknown equations are more dramatic.)

Opening Discussion

(Optional: You might want to briefly review inequality symbols, which will be used in the phone-plan description. Write on the board: >, <, =. Ask volunteers to come to the board to read and write statements using one of the three symbols. Students share ways they remember the direction of the "is greater than" and "is less than" signs.)

Open by displaying a sketch of the salary graphs from *Lesson 8*.

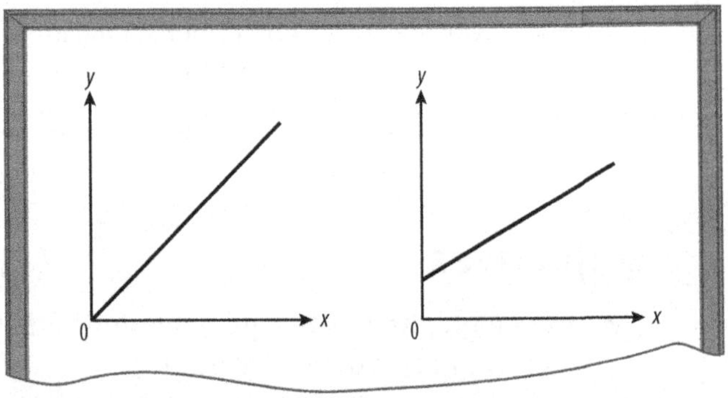

Ask:

What on the graph told you about the starting bonus?

What on the graph told you who was getting a higher weekly rate of pay?

Say:

Today you will have four phone plans to compare in order to make some consumer decisions.

Activity 1: Phone Plans

The instructions below were written for the original, but can be adapted for the new version. Pair up students and distribute an envelope of cards to each. Ask for a volunteer to read aloud the directions for *Phone Plans* (*Student Book*, p. 114). Allow everyone to work independently for 10 minutes, attaching the graphs to their matching representations. Observe which features draw students' attention and are used to help them connect representations.

Challenge those who finish quickly with an additional question:

Which monthly plan is the best for $50? Why?

Then ask students to pair up again to share their reasoning. Ask some pairs that come to agreement to write the equation, table, and ad for one plan on the enlarged versions of the graphs that you prepared earlier.

Pose questions to the class that illuminate the graph features:

What tells you these go together?

Where do you see the table data in the graph?

Where do you see the equation in the graph—the coefficient, or the constant number added?

Invite students to the board to demonstrate how they made the connections. In particular ask:

Phone Plan A

Why does this line start here? (Point to the origin.)

Phone Plan B

Why does this line start here? (Point to (0, 5).)

Which graph is steeper—Plan A's or Plan B's? What does that tell you?

Phone Plan C

How is this graph different from those for Plans A and B?

Why is part of the graph flat? (Encourage references to the tables and equations to support statements made.)

What is happening at this point? (Point to (1,000, 40).)

What would the ad say if the graph looked like this? (Draw a flat line: $y = 40$.)

Phone Plan D

Compare the graphs of Plan D and Plan C.

Clear up any confusion about the equations by asking questions such as

Why do you subtract 1,000 from *M* in the equation for Plan C and subtract 250 from *M* in the equation for Plan D?

Ask for some examples to emphasize the connection between the situation described verbally in the ad and symbolically in the equation. Then ask about the rate of change:

Which plan shows the fastest rate of increase in cost for any period of time? What tells you that information?

Expect to hear that some see the rate of increase in the graph and others in the equation for Plan C.

Heads Up!

The conversation in this first activity can get very involved. Move on to the next activity. In the summary, continue discussing connections between representations.

Activity 2: It Would Depend on the Person

The instructions below were written for the original, but can be adapted for the new version.

Turn to *It Would Depend on the Person* (*Student Book*, p. 116).

This activity uses the phone-plan representations to support a consumer decision. Students will need all four re-pieced ads to come to a good recommendation. Three kinds of customers are portrayed.

Suggest students count off: 1, 2, 3, 1, 2, 3 … to determine the number of the problem they will work on. Allow them to work independently for about 15 minutes.

When most students have completed the problem, ask all those with the same numbers to form groups to share their answers and the ways they made their decisions. Allow time for each group to come to a common agreement for a phone plan for its customer. Then ask a spokesperson from each group to make a persuasive recommendation to their customer as to the *best* and the *worst* plan for that person. Every statement should be justified using at least one type of representation.

If the following have not been answered by the group's spokesperson, ask:

Which representation(s) did you use to help you decide?

What information in that representation swayed your decision?

How is your decision supported by another representation?

Assign Problem 4 to groups who finish early or for homework.

Summary Discussion

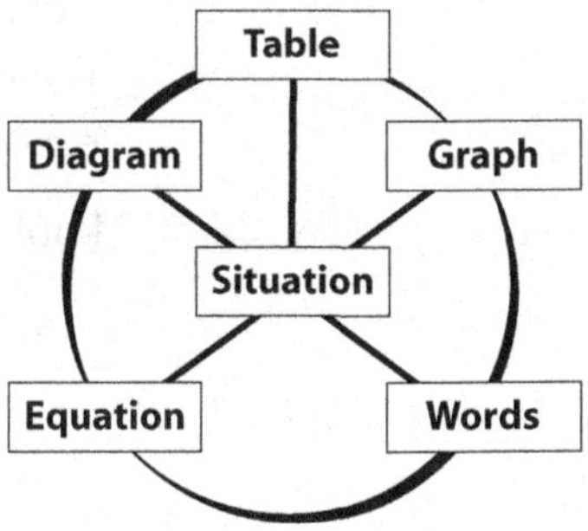

Discuss the different representations everyone used to help make decisions and what information was gleaned from each representation. Ask:

Which type of representation would you turn to first if you had to make this kind of decision again? Why?

Encourage specificity in responses, using today's work for points of reference. Refer students to their *Reflections* section (*Student Book*, p. 162).

Practice

I Am Changing the Rules!, p. 117
Gives students an opportunity to adjust one of the plans and its representations accordingly.

Symbol Sense Practice

Greater Than, Less Than, p. 119
Asks students to express conditions in mathematical notation using inequality symbols.

Solving Two-Step Equations, p. 120
Suggests some ways to solve equations in the form $ax \pm b = c$, using a simple strategy.

Extension

Looking at Four Graphs, p. 122
Asks for justification based on intersections of the four phone-plan graphs.

Test Practice

Test Practice, p. 123

Looking Closely

Observe whether students are able to

Match graphs, tables, equations, and verbal rules by identifying the related features in each representation

If students gravitate toward certain numbers and make decisions based on them—such as the number 1,000, which surfaces in the equation, the table, the graph, and the ad for Plan C—ask where the number occurs in other representations. What explains the flatness of this line? What tells you the graph starts on the y-axis?

Look for informal connections that arise, such as

- When Out numbers increase in a table as the In numbers increase, the line goes up to the right.
- When Out numbers stay the same in a table as the In number increases, the line is flat.
- The number multiplying x in an equation is the difference between the y numbers in the table.
- When the graph does not start at the origin, a number is added to another in the equation.
- If one line is steeper than another, the cost is increasing at a faster rate in the first line.

Connect the flatness of a horizontal line on a graph to a situation in which there is no change over time

Label a few points on the graph, and connect these to numbers in the tables and the ads, if students do not understand what the flatness indicates.

Ask students to think of other situations where there is no change over time (flat rates). For instance, if a heart-monitor graph shows a **flat line**, what does that mean?

Use information from tables, graphs, rules, and equations to support consumer decisions

Ask students what each person would be looking for in a phone plan. Students should be able to explain their decisions by referencing information presented in the various formats.

WHAT TO LOOK FOR IN *LESSON 9*	WHO STANDS OUT? (LIST STUDENTS' INITIALS)			NOTES FOR NEXT STEPS
	STRONG	ADEQUATE	NEEDS WORK	
Representations Graphs • Explains reasons for shape of each graph				
Connections • Matches representations with confidence • Flexibly moves among representations, connecting key features • Explains how table information shows up on each graph • Explains how features of the equations connect to the graph and the table				
Problem Solving • Uses information in the representations to solve problems and support conclusions				

Rationale

This lesson focuses attention on the different representations of a consumer scenario, and the capacity of algebra to simplify complicated situations. Learning to use the tools of algebra efficiently helps students evaluate situations that involve more than one set of conditions.

Math Background

In this lesson, the slant of a constant function, which appears in a graph as a flat line, is considered (a constant function with zero slope). In more formal algebra, the disappearance of x in such situations is explained by the fact that $y = mx + b$, and $m = 0$. Here, however, students focus on the constancy of the output, no matter the value of the input.

Notation is introduced for a situation where the rule changes at a certain point. [When $M \approx 1{,}000$, $C = 40$. When $M > 1{,}000$, $C = 40 + 0.35\ (M - 1{,}000)$].

Context

Phone plans are evolving, and many people have to make decisions about cell-phone and long-distance plans. The pricing structure for phone plans used in this lesson was common in 2004, and updated in 2022 (see *Blackline Master 6a* pages). You might want to consider other pricing scenarios that have evolved since.

Facilitation

Making the Lesson Easier

Start with matching the graphs and tables first. Then match the ads with the equations.

Making the Lesson Harder

Change aspects of the graph, and ask people to predict how the changes might affect the equation and vice versa.

Gather some advertisements, such as grocery ads from three stores, and ask students to develop representations for one product.

Activity 2: It Would Depend on the Person motivated students to bring their everyday decision-making skills to bear when choosing the phone plan best suited to each person's particular phone habits.

For the final question on their own best phone plan, some students were sure of their monthly usage—400 minutes, for example—and compared plans accordingly:

A = $ 28 per 400
B = $ 30 per 400
C = $ 30 per 400
D = $ 40 per 400

The flat rate of $40.00 for 1,000 minutes appealed to many:

Plan C will be better for me, I make call anytime and 1000 minutes is a lot of minute at the rate of $40- per month.

Barbara Tyndall used a two-part rubric ("What I notice" and "My interpretation and next steps" to help her students examine their work. Below are examples of two students' written work from Practice: I Am Changing the Rules!, along with Barbara's written comments.

What I notice:

- Can write an equation to go with the verbal description.
- Creates a table that is accurate and reflects the initial cost.
- Correctly graphs the equation for the new rule, new equation, new table!

My interpretation and next steps:

- Can be alerted to correct notation ($0.07 or 7¢, not $0.07¢).
- Could put "0" in minutes column for first entry (student had erased the zero).
- In response to my asking what the student notices about the two line graphs, he or she mentions "constant rate of change" and "slope."
- Student is ready to go to next lesson!

Student 1 Response to I Am Changing the Rules!

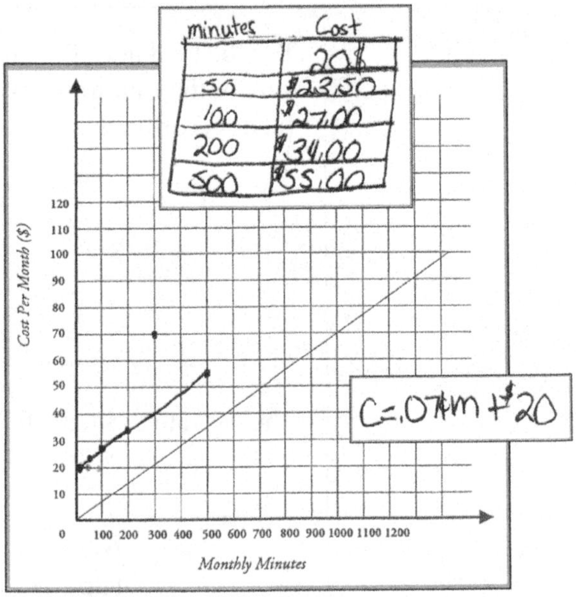

What I notice:

- Writes a new equation that shows the new rule.
- Correctly calculates the cost for 50 minutes, but subsequent costs are incorrect in the new table.
- Graph starts out at the correct point (0, 20), but does not accurately reflect the data in the table.

My interpretation and next steps:

- Understands the connection between the verbal description and the equation.
- Can use the equation to calculate the cost between the first value (of 50 min.) for time, but seems to use a different pattern to complete the table.
- May have difficulty estimating the values on the scale.
- Needs more practice before going to the next lesson.

Student 2 Response to I Am Changing the Rules!

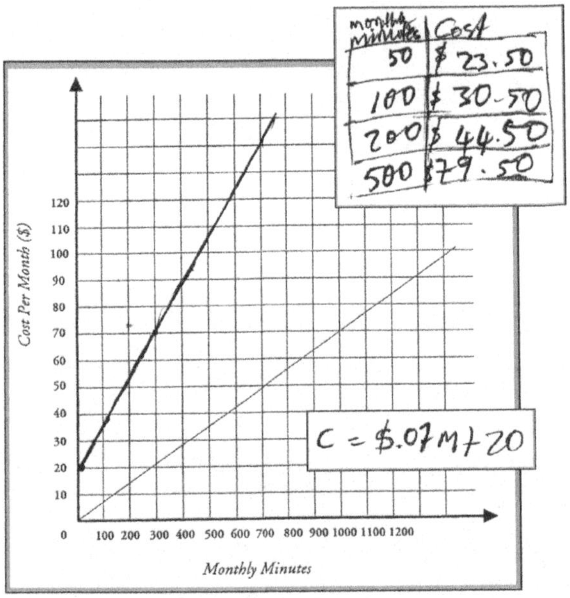

- Should be urged to check that the values in the graph all follow the rule described.

Lancaster-Lebanon Intermediate Unit/13 Career Link, Lancaster, PA

FACILITATING LESSON 10

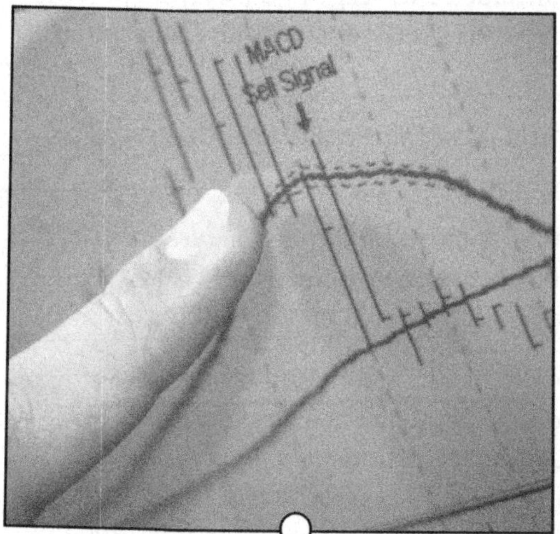

Signs of Change

How do I know which one is changing fastest?

Synopsis

Earlier lessons alluded to the rate of change, as students compared slants of graphs and discussed where change was happening fastest. Here students investigate evidence of and compare constant rates of change, or slope, in linear relationships.

1. The term "linear relationship" is introduced in reference to previous work.
2. Student pairs work on *Activity 1: Seeing the Constant Rate of Change*, and the class discusses constant rate of change in the context of an office assistant's accumulated wages.
4. Student pairs rank and justify the rates of change for six representations—two equations, two tables, and two graphs. They propose situations that might correspond to the data in these multiple representations.
5. The class reaches agreement on rankings and connects constant rate of change to the slope number.

Objectives

- Locate indicators of a constant rate of change in graphs, tables, and equations of linear relationships
- Compare rates of change in a variety of linear representations

Materials/Prep

- Colored markers
- Easel paper

Reproduce on easel paper or presentation slides a variety of graphs from previous lessons, including graphs where y increases, decreases, and remains constant as x increases.

Enlarge the graph for *Seeing the Constant Rate of Change* (*Student Book*, p. 126) for later posting.

Make copies of the equations, tables, and graphs on *Equations, Tables, and Graphs, Blackline Master 7,* and cut each into six cards to distribute for *Activity 2.* These cards will be easy to manipulate as students decide upon rankings.

A note on vocabulary: A **linear relationship** or linear function is one whose ordered pairs, when graphed on a coordinate plane, form a straight line. The **constant rate of change** is a feature of all linear relationships. Every time the x-value increases by a given amount, the y-value rises or falls by a constant amount. The **slope** is the steepness of the slant of a line on a graph, quantified by how much it rises or falls for each unit the line moves to the right.

Opening Discussion

Begin by displaying a variety graphs from previous lessons.

Ask:

What do these graphs have in common?

Students should mention that all the graphs are straight lines; they all compare two variables; and they all can be represented as tables, a rule in words, and equations.

Point out that this unit has focused on seeking patterns and building rules to describe the relationship between two variables. Explain that the graphs of all these relationships are straight lines, and relationships that graph as straight lines are called linear relationships.

Say:

Today you will look at an important aspect of these linear relationships: the rate of change, or the rate at which variables increase or decrease in a relationship.

Activity 1: Seeing the Constant Rate of Change

Suggest students turn to *Seeing the Constant Rate of Change* (*Student Book*, p. 126), with the graph "An Office Assistant's Wages Over Time." Review directions, then pair students to complete the exercise.

When everyone has finished at least Questions 1–3, draw the class together. Focus attention on the line and axes by asking:

As you increase the *x*-value by one, what happens to the *y*-value?

As volunteers tell what they see happening, mark the "steps" in the enlarged graph posted for all to see. Say, as you do:

For every time the number of hours goes up by one, the earnings go up by 15.

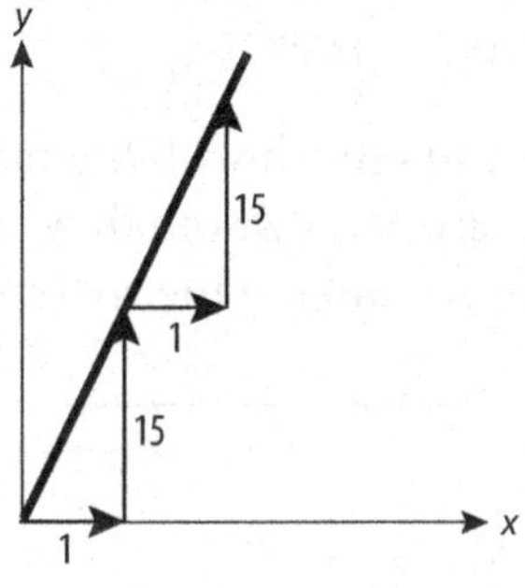

Mark three steps on the graph, and encourage students to draw in the steps on their graphs. Ask a volunteer to mark the rest of the steps on the graph you posted. Label the steps with the numbers "1" and "15" to emphasize the constant rate of change. Then ask:

What do you notice about the rate of change as we have marked it here?

When it seems clear that students see a pattern that builds or grows steadily by $15 for every one-hour increase, say:

Mathematicians call this type of change a "constant rate of change." Why do you think they describe the change as constant?

If no one mentions it, raise the point that for equal intervals between *x*-values, the change on the *y*-axis is constant, or the same.

Ask students to state in their own words what the term "constant rate of change" means, and post their definitions on the class vocabulary list.

Allow time to discuss the *What If* questions (Questions 4–6, *Student Book*, p. 127). Bring up the following points if no one else does:

- The constant rate of change affects the steepness of the line; the higher the coefficient, the steeper the line.
- A bonus, or amount added to an equation, does not affect the steepness of the line (its rate of change).
- The slant of a graph gets steeper when the constant rate of change increases. When the constant rate of change decreases (the coefficient is a smaller number), the graph becomes flatter.

Activity 2: Ranking Rates of Change

Tell the class:

Now you will have a chance to apply what you know about constant rate of change.

Regroup pairs and review directions for *Ranking Rates of Change* (*Student Book*, p. 128). Distribute envelopes with cards of graphs, tables, and equations (*Blackline Master 7*).

When pairs finish with the rankings, encourage them to find another pair of students with whom they can discuss their rankings. When everyone has completed ranking the six representations, bring the class together.

Heads Up!

The rankings are (1)—Table C; (2)—Graph F; (3)—Equation A; (4) and (5) are the same—Table D and Graph E; (6)—Equation B.

Ask a few pairs to post their rankings, and invite comments from the class. If there are disagreements, ask for justifications.

When students reach consensus on all six rankings, summarize the ways you saw people making decisions and pose the following general questions:

Where do you find the rate of change in equations, tables, or graphs?

How does changing the *y*-intercept change the graph or the table?

Parallel lines never intersect. What conditions cause lines to be parallel on a graph?

The last question may prompt the need to graph both representations with the "5" coefficient—Graph E and Table D.

Take some time to focus on Table D.

Ask:

Where can you see that for every time *x* increases by 1, *y* increases by five?

This is most clearly evident in the change between $x = 10$ and $x = 11$. Encourage students to look for it elsewhere. Can they detect the "over 1, up 5" elsewhere?

Some might see that as *x* increases by 4 from "3" to "7," *y* increases by 20 (from 65 to 85), which is the same as the five-to-one ratio. If students have studied equal ratios, connect $5/1 = 20/4 = 15/3$.

Table D

In (x)	Out (y)
3	65
7	85
10	100
11	105

x increases by 4 — y increases by 20

x increases by 1 — y increases by 5

Connecting to a Situation

Redirect attention to Question 3, which asks for a real-life situation that would connect to one of the representations.

Ask two or three students to write their stories on the board, and challenge the class:

Which representations go with each of these stories?

Students should justify their answers to show that the representations and stories connect for all cases.

Summary Discussion

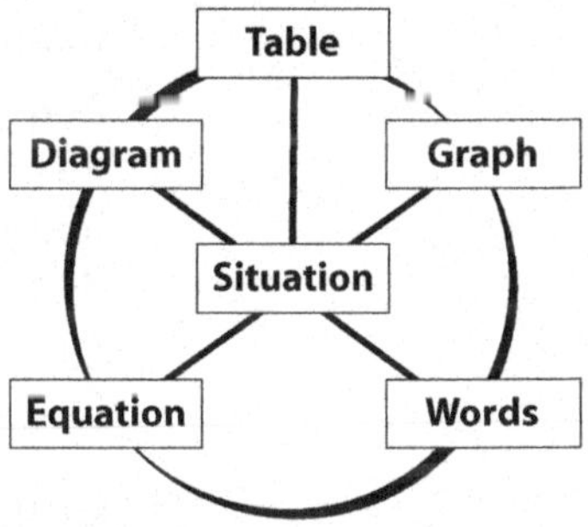

Close the discussion by introducing the term "slope." Tell students they have begun to see more deeply into linear patterns today because they saw that in addition to the relationship between x and y, every linear pattern has a constant rate of change. That constant rate of change is called the slope, and that slope can always be described with numbers. Ask:

What does the rate of change tell you?

Where is the number that indicates slope in the equation? In the graph? In the table?

Say:

In the next lesson, you will see that not all changes are constant and not all patterns are linear.

Practice

Watching Money Grow in a Table, p. 129
Investigates three different rates of change for CD investments.

Whose Total Earnings Change the Fastest?, p. 130
Asks students to examine and compare the rates of change for three graphs.

Symbol Sense Practice

Solving More Two-Step Equations, p. 132

For practice solving equations in the form of $ax \pm b = c$, and $x/a \pm b = c$.

Test Practice

Test Practice, p. 133

Looking Closely

Observe whether students are able to

Locate indicators of constant rate of change in graphs, tables, and equations of linear relationships

Looking for rate-of-change patterns is different from looking for the pattern that describes the relationship between the two variables. If students are having trouble distinguishing between these patterns, start with the graph. Label some consecutive points. Clarify that you are not looking for the relationship between the x- and y-values (which gives the rule); rather you are looking at how y changes as you move from point to point in even increments of x.

Then, referring to the table, suggest to struggling students that looking down the y column will help them see the rate of change. When they compare the x-value to the y-value, the rule emerges. But when they look at the differences in the y-values as the x-value increases steadily, they will see the rate of change that is an important part of the rule—it is the coefficient of x! Look at both tables to connect the changes in consecutive y-values with the respective coefficients, drawing attention to the relationship.

Compare rates of change in a variety of linear representations

Always start with the visual. Compare the lines to ramp inclines. When the rate of change is larger, the slope of the line is steeper because y increases significantly every time you move up a given interval on the x-axis. Conversely, the rate of change is smaller when you move over the same interval on the x-axis but hardly move up at all on the y-axis.

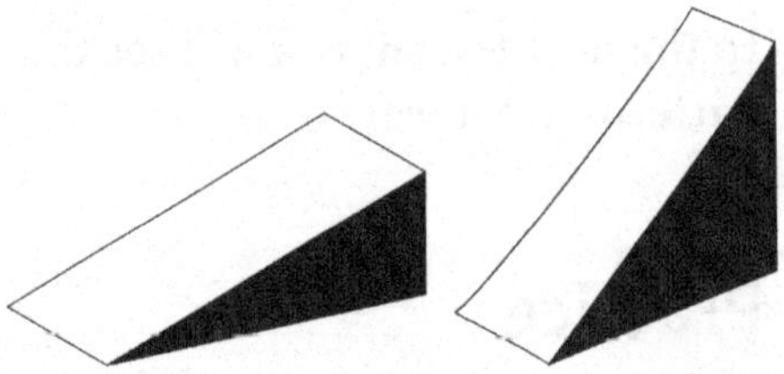

As students make the rate of change connection to the equation, they might understand more clearly that any rule involving multiplication by 25 will produce a faster constant rate of change than any rule involving multiplication by 10.

WHAT TO LOOK FOR IN *LESSON 10*	WHO STANDS OUT? (LIST STUDENTS' INITIALS)			NOTES FOR NEXT STEPS
	STRONG	ADEQUATE	NEEDS WORK	
Representations Graphs • Locates and quantifies the constant rate of change in a linear graph Tables • Locates and quantifies the constant rate of change in the tabular representation of a linear pattern Rules • Locates and quantifies the constant rate of change in the graphical representation of a linear pattern				
Connections • Connects the constant rate of change in the various representations				

Rationale

This lesson provides an opportunity to use what students know about algebraic tools—graphs, tables, and equations—to gain a fuller understanding of what constitutes a linear pattern. Constant rate of change is a defining characteristic of linear patterns, and it is one characteristic used to discriminate between linear and nonlinear patterns, as will be done in *Lessons 11* and *12*. Not only is this knowledge useful in the continuing study of mathematics, it is also useful for quick, graphical comparison of data to determine which pattern is growing faster. Such comparisons can aid in analysis and decision-making in many fields of work.

Math Background

Slope is a key concept in the study of linear patterns. The treatment in this lesson anticipates study of the formal and general slope formula:

$$m = \frac{(y_2 - y_1)}{(x_2 - x_1)}$$

However, that formality is foregone here. Rather, the emphasis is on seeing the slope as a number that appears in a particular location in each of the representations. Locating and quantifying the slope here is a way to help students make sense of its meaning before rushing to the general formula.

Certainly, if some students remember the slope formula from algebra classes of the past, ask them to make the connection between the points on the graph and entries in the table.

Facilitation

Making the Lesson Easier

Give more examples of graphs and tables.

Making the Lesson Harder

Move the class toward an understanding of the slope formula:

$$m = \frac{(y_2 - y_1)}{(x_2 - x_1)}$$

Ask that they consider how they found the changes in the table. Focus on where the subtraction comes from and how division factors in. Use this knowledge to connect the slope formula to the way rate of change is depicted in the table. Use the formula to calculate the rate of change, or slope, for the graphs in today's lesson.

The idea of rate of change—how it looks on a graph and how it is seen in a table or equation—takes time to learn. Here the teacher has been comparing In-Out tables and asks students to imagine what the graphs would look like and why. When students appear stumped, she moves back to concretely representing the graph.

Teacher: "For each time your *x* goes up one, how much does your *y* increase?"

Student: "By five."

Teacher: "What happens if *x* goes up another one. What happens with *y*?"

Student: "Same thing."

Teacher: "What is the rate of change?"

Student: "Five."

Teacher: "Another word for this: slope. It tells you how it is increasing."

Teacher: "When you look at the equation, how does it connect with the table?"

Student: "That five it goes up is in the equation."

Table for $y = 5x + 6$

In (*x*)	Out (*y*)
1	11
2	16
3	21
4	26
5	31

The teacher creates another In-Out table, this one with a constant rate of change of two and asks:

Teacher: "So, what would that look like, compared to this one? Can you show me? Is it a bigger rate of change or smaller?"

Student: "Smaller. It's five here and two here."

Teacher: "On a graph, would the line be steeper or shallower?"

Student: "Steeper?"

Teacher: "Does everyone agree?"

(The class is silent.)

Table for $y = 2x + 6$

In (*x*)	Out (*y*)
1	8
2	10
3	12
4	14
5	16

continued on next page

continued from previous page

Teacher: "Let's investigate that question. Let's put it up and see whether it's steeper or shallower."

A student creates the graphs at the board and sees that the line with the rate of change of two is less steep than the line with the rate of change of five.

Teacher: "What do you think? Did that surprise you? Let's look again at what happened between the *x* and the *y*. Go over one and up two. So the slope of the line is two. Slope is another way to look at the rate of change. If you were walking up hill, which would be steeper?"

Solange Farina
BEGIN Managed Programs, New York City

FACILITATING LESSON 11

Rising Gas Prices

Are gas prices rising steadily?

Synopsis

In order to see that not all growth patterns are linear, **nonlinear patterns** are introduced for comparison. The first activity returns to a visual problem similar to the banquet tables problem of *Lesson 2*, and the second is a comparison of two gas-pricing scenarios. The focus of both activities is examining the evidence for rate of change (constant and nonconstant) in diagrams, tables, graphs, and equations.

1. Pairs examine the rates of change in two visual growing patterns and share their findings with the class.
2. Pairs compare two pricing scenarios, supporting their reasoning with tables and graphs, and then they share their conclusions.
3. The class summarizes by classifying different types of patterns as showing constant or nonconstant rates of change.

Objectives

- Produce and examine tables for linear and nonlinear patterns to answer the question: Are values changing at a constant rate?
- Produce and examine graphs for linear and nonlinear patterns to answer the question: Does this depict a constant rate of change?
- Predict for specific cases, using derived rules

Materials/Prep

- Calculators
- Grid paper
- Easel paper (preferably with grids)
- Rulers

Opening Discussion

Begin by asking students to review what they have learned about constant rate of change: For example, the graph is linear, the increases in the table are equal, and the coefficient of x in the equation gives information about the magnitude of the constant rate of change. Then say:

Today you will see that not all patterns change at a constant rate. Different types of patterns vary in the way they look in tables and graphs. By the end of class, you should be able to distinguish quickly at least two types of change that occur in patterns.

Activity 1: Patterns of Change

Ask pairs to examine the patterns on p. 136 of the *Student Book*. Say:

Study the patterns, and make some notes about how each changes. Then extend each pattern to examine how the patterns continue to grow.

Record observations made for each pattern. Students might say:

- You add 4 successively in Pattern 1.
- You add 3, then 5, then 7, and so on in Pattern 2.
- The difference in what is added is 2 in Pattern 2.
- Pattern 1 is changing at a constant rate, but Pattern 2 is not.

Ask pairs to predict for the 10th and 25th cases and to establish the general rule for any case. Encourage the use of all tools to arrive at the equations.

Call the class together to discuss findings for each pattern.

In the first pattern, the change is constant: Four dots are added in each instance. Ask:

Where do you see that change in the diagrams? In the table? Where do you see that change in the graph? Where is it in the equation?

When the idea of changing at a constant rate is established as "every time you go up one shape, you add or subtract the same amount, four," focus on the second pattern.

In Pattern 2, the pattern grows, but the change happens very differently. A different amount is added each time; therefore, the change is not constant. Ask:

Where do you see that change in the diagrams? In the table? Where do you see that change in the graph? Where is it in the equation?

Heads Up!

The second pattern presents an opportunity to introduce the **exponent** as notation for multiplying a number repeatedly by itself. If it has not arisen in the discussion, write:

1 x 1, 2 x 2, 3 x 3, 4 x 4, ... as 1^2, 2^2, 3^2, 4^2.

Read these as "one squared, two squared, etc." For practice using exponents, assign *Symbol Sense Practice: Squaring Numbers* and *Test Practice (Student Book*, pp.144 and 146.)

Then introduce the more general use of the exponent:

2 x 2 x 2 can be written as $2^3 = 8$

$2 \times 2 \times 2 \times 2 = 2^4 = 16$

$2 \times 2 \times 2 \times 2 \times 2 = 2^5 = 32$

Assign *Symbol Sense Practice: Using Exponents* (*Student Book*, p. 145).

Explore this with the calculator as well.

Activity 2: Sky-High Gas Prices

Allow time for student pairs to work together on *Activity 2* (*Student Book*, p. 138).

Ask some pairs to post their tables and graphs. Tell them to provide evidence of the change in the graphs, encouraging them to draw the steps to show the change. Emphasize how the constant slope can be seen clearly on the straight line (at Joe's), whereas the Metric slope increases from point to point.

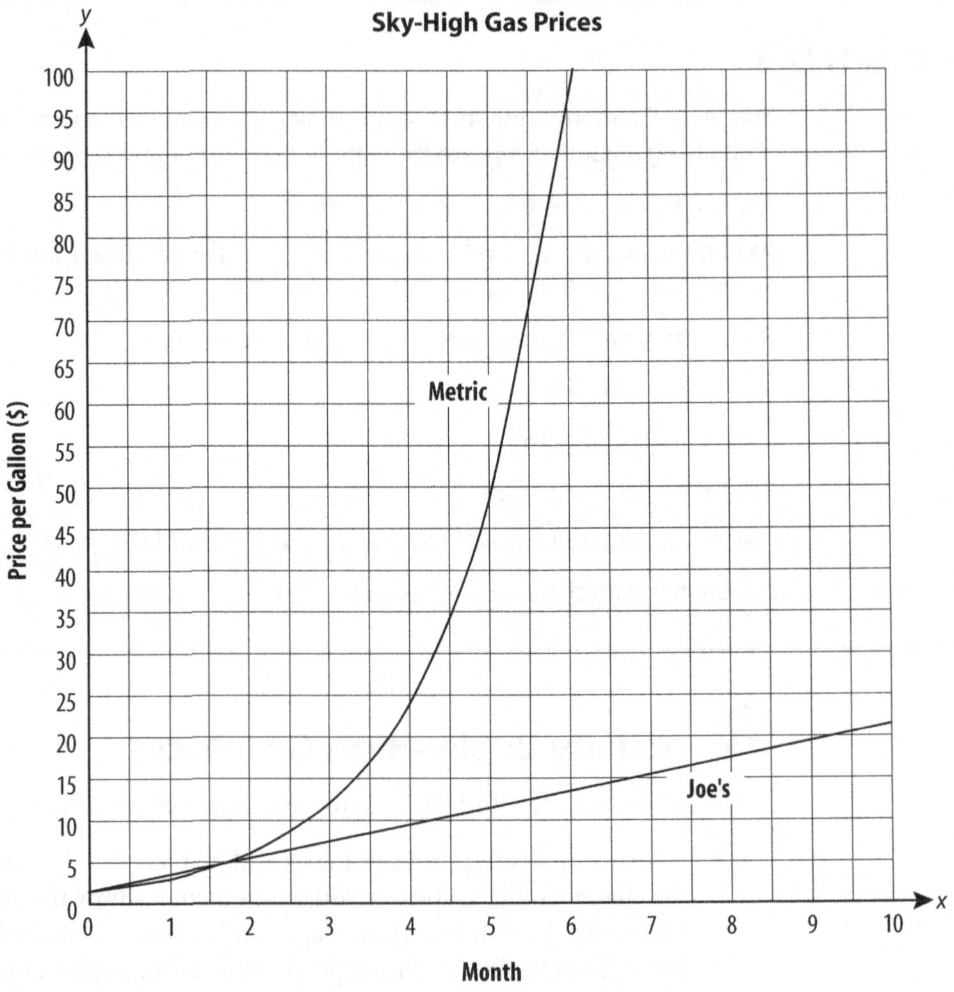

Discuss what made students suspect that the price of gas at Metric would quickly become more expensive than the price of gas at Joe's. Compare predictions for the prices at the end of the year. Ask:

Which tool helped you figure out the gas prices at the end of the year?

Focus on the equations only if you think students are ready. Students should be able to write the general rule for Joe's:

$y = 2x + 1.50$, a linear equation.

However, the notation for the second (exponential) equation may be too unfamiliar at this point, so limit the discussion to a comparison of the tables and graphs. If you think students are ready to make sense of it, opt to introduce the equation: $y = 1.50(2)^x$. Ask:

How does this equation explain the general rule for Metric?

How could you use this equation to calculate the price of a gallon of gas after three months? [$\$1.50(2)^3 = \$1.50(2)(2)(2) = \$1.50(8) = \$12.00$]

Six months? [$\$1.50(2)^6 = \$1.50(2)(2)(2)(2)(2)(2) = \$1.50(64) = \$96.00$]

Explain that mathematicians refer to this pattern of growth as **exponential** growth.

Summary Discussion

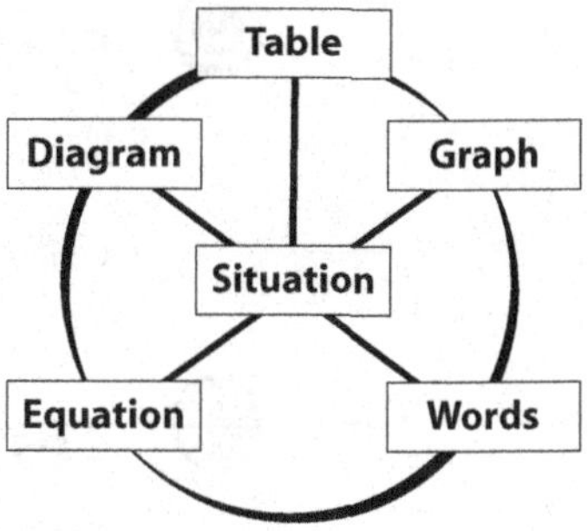

To solidify today's learning, ask a volunteer to summarize.

We looked at two kinds of patterns today. Who can summarize by describing those two patterns?

Continue the discussion until the following points about linear and nonlinear patterns are raised and posted on the board:

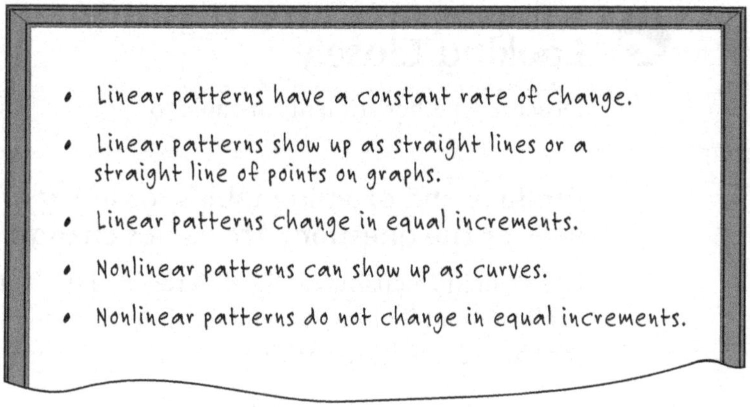

Provide a few moments for students to compose a definition for "nonlinear patterns" and post it on the class vocabulary list. Students record the definition in *Vocabulary* (*Student Book*, p. 156).

Close with time for students to write in *Reflections* (*Student Book*, p. 162).

Practice

More Patterns of Change, p. 140
Asks students to check for linearity and nonlinearity of visual growth patterns.

Investigation A—Expanding Squares, p. 141
Compares the rates at which perimeters and areas of squares grow as the length of the side increases.

Investigation B—What Is Behind the Door?, p. 142
Asks students to choose the better deal in a game show.

Investigation C—What Size Is That Copy?, p. 143
Challenges students to examine the effect on dimension and area of a dollar bill when it is enlarged and reduced on a copy machine.

Symbol Sense Practice

Squaring Numbers, p. 144

Using Exponents, p. 145

Test Practice

Test Practice, p. 146

Looking Closely

Observe whether students are able to

Produce and examine tables for linear and nonlinear patterns to answer the question: Are values changing at a constant rate?

Are students focusing on the difference in the values of y as x increases by one? Be explicit about looking down the columns from y to y, rather than looking across the rows, comparing y with x.

Produce and examine graphs for linear and nonlinear patterns to answer the question: Does this depict a constant rate of change?

Are students focusing on the step patterns as well as the straightness of the line as cues for constant rate of change? Draw in the steps carefully, and connect the points either with a straight line or a curve. Label steps with numbers to make the rate of change more evident.

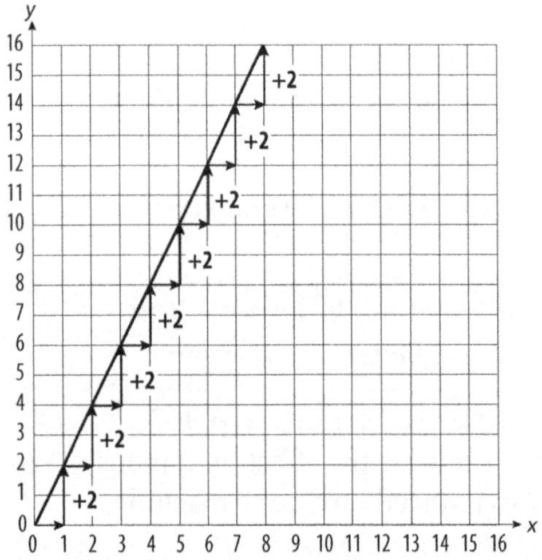

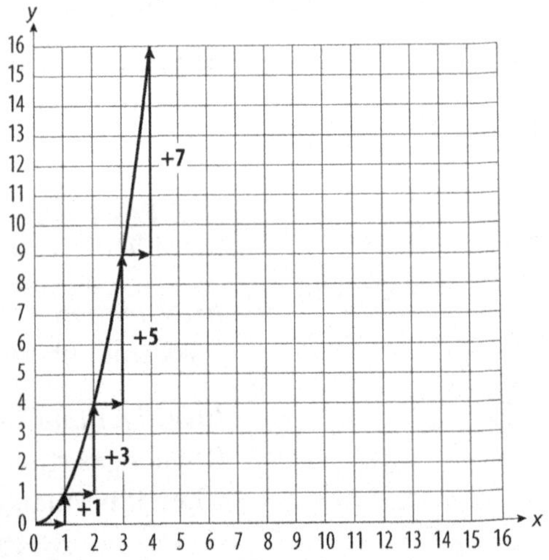

Predict for specific cases, using derived rules

Can students confidently predict for the 10th and 25th cases for the linear situation (Joe's)? Are they able to express and use the general rule in algebraic symbols?

In the exponential situation (Metric), can students use a general rule? How do students predict from the rule? Can they express the fact that the cost doubles over and over again? Are they able to connect this to the fact that the number of doublings (multiplications by a factor of two) is connected to the number of months? If you have introduced exponential notation, can they connect to and predict from the equation: $y = 1.50(2)^x$?

WHAT TO LOOK FOR IN *LESSON 11*	WHO STANDS OUT? (LIST STUDENTS' INITIALS)			NOTES FOR NEXT STEPS
	STRONG	ADEQUATE	NEEDS WORK	
Representations Diagrams/Models • Accurately diagrams the patterns Tables • Uses table data to explain different types of rate of change (constant versus nonconstant) Graphs • Uses the graph to explain the differences in the rates of change (constant versus nonconstant) Rules • Accurately writes the relationship between the case numbers and the total number of dots • Accurately describes relationship (in words) between the number of months and the price of gas				
Connections • Can articulate characteristics of a linear pattern • Can articulate the difference between a linear and a nonlinear pattern				
Problem Solving • Uses algebraic tools to solve the problem accurately				

Rationale

The National Council of Teachers of Mathematics' (NCTM's) Algebra Standard for Grades 6–8 says, "All students should be able to identify functions as linear or nonlinear and contrast their properties from tables, graphs, and rules." Recognizing these differences helps students understand the various ways in which patterns can grow and change, and prevents the misconception that all patterns are linear. Linear relationships remain the focus of this unit, but we introduce exponential and **quadratic patterns** as a way to highlight the special qualities of linear relationships.

Math Background

Exponential and quadratic relationships clearly differ from linear relationships in their tabular, graphical, and symbolic representations. In the activities, students notice that the rate of change, or difference between consecutive terms, is repeated in linear relationships, but in other relationships it is not.

Quadratic Relationships

In a quadratic function, the equation contains squared terms, and two is the highest power to which any term is raised—for example, $y = x^2 + 5$. The general form of a quadratic function is $y = ax^2 + bx + c$. In the dot squares pattern, the equation is $y = x^2$. The graph is curvilinear. The difference between successive elements is not constant; rather, the change varies, going from three, to five, to seven, and so on.

Rate of Change in the Quadratic Function $y = x^2$

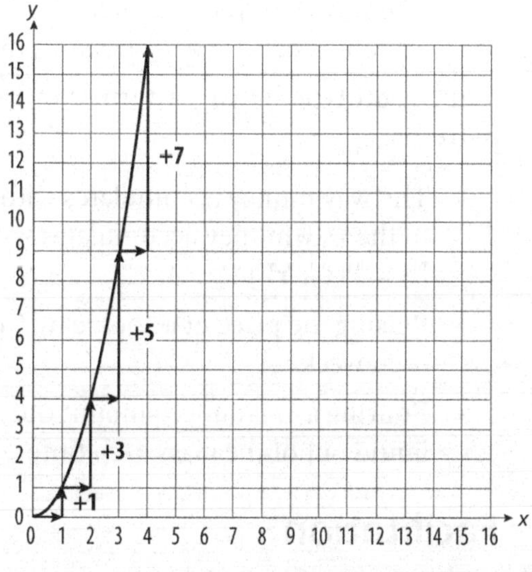

Exponential Relationships

In exponential functions the equation contains the variable as the power to which a constant is raised—for example, $y = 2^x$. The graph is curvilinear, and the rate of change varies in that it is successively multiplied by the constant. In the gas station scenario, $y = \$1.50(2)^x$, the difference in the price of a gallon of gas doubles every month.

Rate of Change in the Exponential Function $y = 2^x$

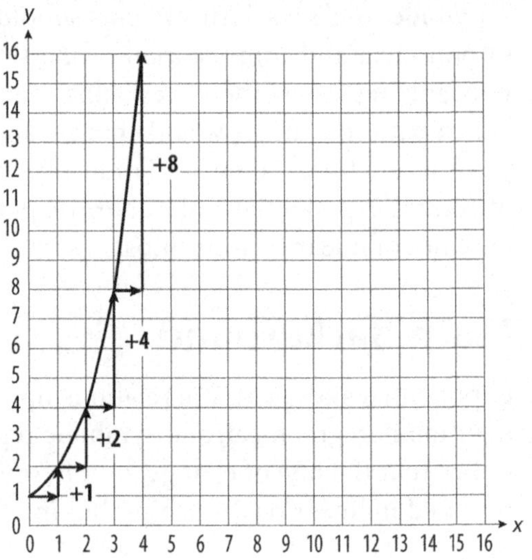

Context

There are several scenarios that highlight the differences between linear and nonlinear situations. Explore some with students to strengthen the comparison.

Some typical linear patterns contrasted with quadratic patterns:

- The way perimeter changes as the length of the side changes ($P = 4s$) contrasted with the way the area changes ($A = s^2$)
- The way the circumference of a circle ($C = 2\pi r$) grows contrasted with the way the area ($A = \pi r^2$) grows as the radius increases

Some prototype linear patterns contrasted with exponential growth or decay patterns:

- The way money accumulates with simple interest [$A = P(1 + r)t$] in contrast to the way money accumulates with interest compounded annually [$A = P(1 + r)^t$]
- Raising the price of a cookie by 5 cents per week compared to raising it 5% each week
- Pouring a 1/2-ounce amount out of a glass compared to pouring half the liquid out of the glass each time

Facilitation

Making the Lesson Easier

Concentrate on the table and graph representations. Do not focus on the exponential equations.

Making the Lesson Harder

Focus more heavily on the introduction of exponential notation. If students readily see the connections between the table and the graph, ask them to connect these to some equations as well. Compare the linear equations to the ones that use exponents.

For the dot patterns, ask students where they see the two in the equation $y = x^2$. Contrast this with a new dot situation where $y = 2x$. Point out that in the former, the two has nothing to do with doubling.

For the gas station scenario, ask where each element of the equation $y = 1.50(2)^x$ occurs in the table.

Assign the practices (*Investigations A–C*).

This teacher's careful attention to students' explanations and rephrasing of those explanations clarifies "rate of change" for the class.

Teacher: "If I were looking at the equation, what part of the equation tells you about the rate of change?"

Student: "You know that—whatever. Five stays the same; *x* changes. It's five times the *x*. Five stays the same, so that's how you know what you just asked."

Teacher: "That was so nice. You just told me that if the five stays the same, what is the rate? This is a constant rate of change. You told me that *x* can be any number, and *x* always changes. What is the word for that? It's on our list; *x* is a variable in this case. Nice. Now what about the equation $y = 2x + 2$. What's the rate of change?"

Student: "Two."

Teacher: "Where do you find that?"

Student: "The two in front of the *x*."

Teacher: "It's always the number before the *x*. Last week you used the word 'multiplier.' You said it beautifully. That's telling you what the rate of change is."

Solange Farina
BEGIN Managed Programs, New York City

A student and the teacher work together at describing nonlinear change in a graph.

Teacher: "What does the line look like?"

Student: "Curving. It could kind of keep going."

Teacher: "Let's look for a second at a couple of points. If *x* is one, what is *y*?"

Student: "One."

Teacher: "Then if *x* is two, what is *y*?

Student: "Two … four. It increases by two."

The teacher traces some other steps on the graph.

Student: "So each step between points is getting bigger and bigger, so it's not a constant rate of change."

Teacher: "You can call this exponential growth. It's not linear. We just started this today."

Solange Farina
BEGIN Managed Programs, New York City

The choice of scale on the axes significantly affected the appearance of the students' graphs that compare the soaring price at Metric with the steadily rising price at Joe's. Although both these graphs display the information correctly from the table data, in Figure 2, the spread of the increments on the x-axis makes the steadily increasing rate of change on the curve more evident than does Figure 1.

Figure 1

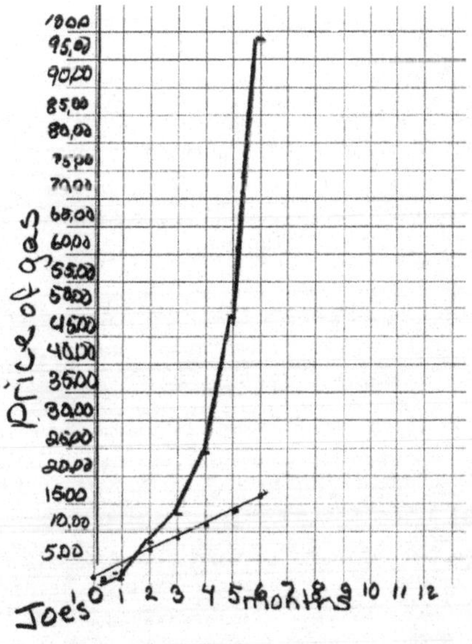

Figure 2

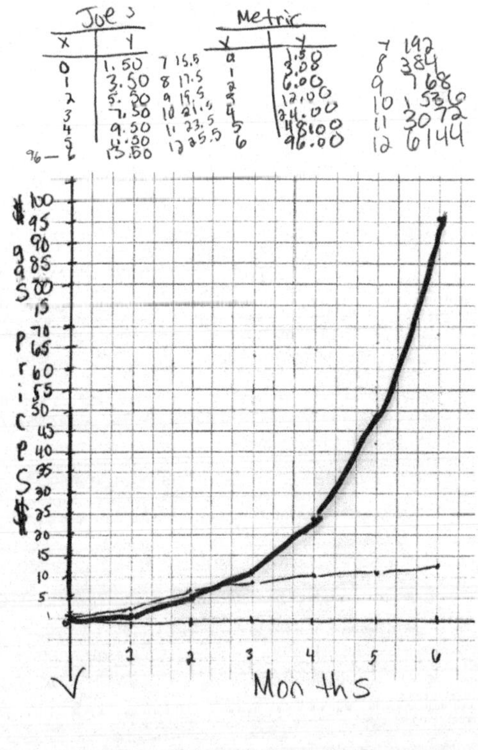

Compilation of student work from pilot classes

FACILITATING LESSON 12

The Patio Project

How can you predict what you will need for the larger design?

Synopsis

This last lesson offers students an opportunity to apply much of what they have learned throughout the unit in an extended investigation that compares three situations: one where the value is constant (no change); a second where the rate of change is a constant positive amount; and a third where the rate of change increases. Students' presentations can also serve as performance assessments.

1. Student pairs solve *The Patio Project* problem and prepare presentations.
2. Students compare presentations and solutions.
3. Everyone examines three "Patio Project" graphs for differences and connections to data.
4. Students assess the problem and their approaches to it.

Objective

- Apply algebraic tools (diagrams, tables, graphs, rules, and equations) to solve a problem

Materials/Prep

- Calculators
- Colored markers
- Grid paper
- Easel paper (preferably with grids)
- Rulers
- Scissors
- Tape

Make copies of *Tiles, Blackline Master 8*, for students to cut out and use. Make two copies for each pair.

Make copies of *Three Types of Tile, Blackline Master 9*, for comparisons in the event that no pair chooses to make graphs as part of their presentation.

Activity: The Patio Project

Review the problem on pp. 148–150 of the *Student Book*. Distribute colored tiles and copies of *Tiles, Blackline Master 8*, for students to cut out and use for modeling. Also distribute one piece of easel paper per student pair to use for presentations to the hypothetical customer. Allow ample time for pairs to work on the problem and to prepare presentations.

Note the tools students choose to work with when solving the problem. Who uses tiles or diagrams? Who develops rules without completing the table?

If students seem stuck, you might ask:

For the next-size patio project, how many corner tiles would you add? How many edge tiles would you add?

How many edge tiles would then be on one side? How do you know?

Is there a relationship between the number of edge tiles on one side and the number of center tiles?

Get a feel for whether students see each case as independent or see the pattern developing. If pairs finish early, invite them to consult with other pairs to share findings.

Have pairs post their presentations for a gallery walk, with students taking time to view several of the presentations. As they examine presentations, ask them:

What similarities and differences in conclusions and solution processes do you see?

Does any presentation raise questions for you?

One point of the activity is to recognize that the number of center tiles grows very rapidly compared to the number of corner tiles, which remains constant at four, and the number of edge tiles, which grows constantly by four. Solicit this observation if no one offers it.

Generate a discussion about variations in change and review where signs of change occur in algebraic representations.

Possible findings are listed below. All solutions, whether or not they look like those listed, benefit from being modeled with diagrams or tiles to clarify and verify.

Corners Tiles

There are four corner tiles.

In general, no matter how large the patio, the number of corner (striped) tiles always is four.

$y = 4$

The graph is a straight horizontal line.

Center Tiles

The number of center tiles = 98 x 98 = 98^2 = 9,604.

More generally, the number of center tiles is the number of tiles on each side less two, squared, $(x - 2)^2$. However, students are more likely to describe this pattern as the number of tiles on a side minus two, times itself or

$(x - 2)(x - 2)$

The graph has a curved shape. The change increases more rapidly as the size of the sides increases.

Edge Tiles

The pattern of edge tiles can be seen in several ways.

Some may see that the number of white tiles is 98 x 4 = 392 or more generally, four times two less than the total number of tiles on a side.

$y = 4(x - 2)$

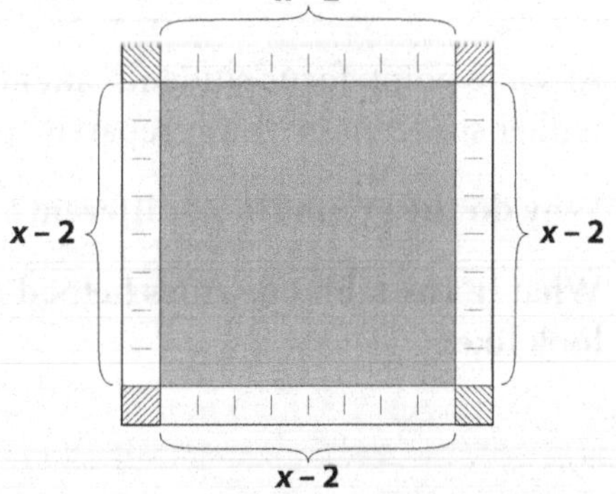

Others might see the number of white tiles is (100 x 4) – 8, having multiplied the number of side tiles by four and then subtracted two tiles per side. More generally

$y = 4x - 8$

Still others may arrive at the expression by seeing the total number of tiles—10,000—and then subtracting the 9,604 center tiles, and then subtracting the 4 corner tiles to arrive at 392. More generally

$y = x^2 - (x - 2)^2 - 4$

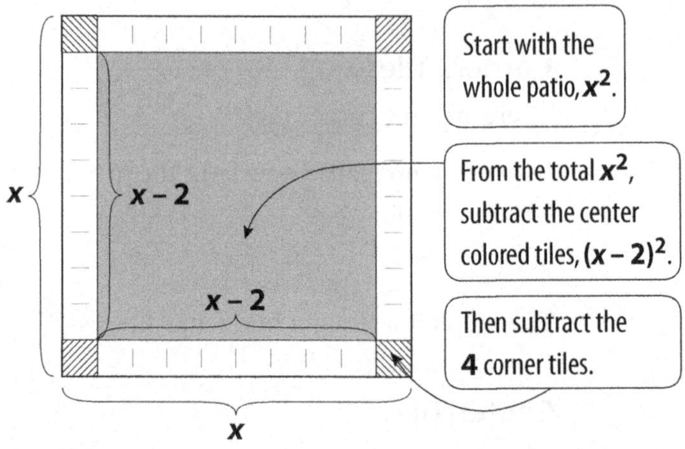

The graph is a straight line. The rate of change is constant (the change in y is four for every change of one in x).

The Total

The total is seen as 100^2. More generally, the total is the product of the side multiplied by the side ($y = x^2$).

The total = 4 striped tiles + 4(98) white tiles + $(98)^2$ solid-color tiles.

In general, $y = 4 + 4(x - 2) + (x - 2)^2$

Heads Up!

The generalization of the total number of tiles as represented by the equation $y = 4 + 4(x - 2) + (x - 2)^2$ is best attempted only in advanced classes.

At some point, focus students' attention on a general comparison of the graphs (either students' or your prepared copies):

Why do the graphs look different?

What in the tables or rules helped you predict what these graphs would look like?

Summary Discussion

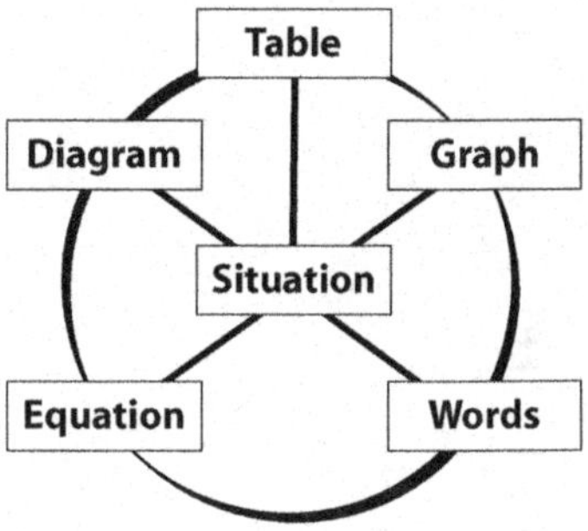

Invite students to assess the problem and their work by asking:

What was hard or easy about this problem?

What helped you solve the problem?

If you had it to do over again, would you change anything about your approach to the problem? What?

Suggest students write about their problem-solving processes in the *Reflections* section of their books, p. 163.

Looking Closely

Observe whether students are able to

Apply algebraic tools (diagrams, tables, graphs, rules, and equations) to solve a problem

Students may select tools they prefer. It is not necessary, for instance, to complete more than three or four entries in the table before starting to look for generalized patterns that can more quickly yield an answer. Some students may find using a graph unnecessary. Many, however, will rely on diagrams or models to see the number of tile types in consecutive patio designs.

If students choose not to work with consecutive patio models, observe how they discern the patterns. If they struggle to see the patterns, suggest they build consecutive models and record that data.

Pay attention to how students solve the problem. If they work with verbal rules, listen carefully. Ask to be shown how they use their rules to predict the next-size model or the 100 x 100 model. In their presentations, the verbal rules should be clearly articulated so you and the other students in the class can make the connection between equations and verbal rules.

WHAT TO LOOK FOR IN *LESSON 12*	WHO STANDS OUT? (LIST STUDENTS' INITIALS)			NOTES FOR NEXT STEPS
	STRONG	ADEQUATE	NEEDS WORK	
Connections • Uses table data to arrive at verbal rules for some or all of the cases • Uses table data to arrive at equations representing some or all of the cases • Makes connections between rates of change and nature of table data, equations, and/or graphs				
Problem solving • Understands problem and enters it appropriately • Solves problem accurately for the 100 x 100 patio				
Presentation • Clearly describes processes used to solve problem, using any or all of the algebraic tools				

Rationale

After directed practice, be sure to let students solve a problem independently. By allowing autonomy, you can more easily assess what students know. You will see which tools they gravitate toward and which provide greater challenges, indicating possible areas for further practice.

Math Background

The work in this lesson builds on earlier work with rates of change. Students apply what they know about connections between table data and equations, for instance, to generalize the patterns and find the number of each type of tile. If they notice that the edge tiles are changing by four in each consecutive case, they may see that the equation must involve four as its coefficient. They may notice that the values for $x = 0$ and $x = 1$ are not practical, since the smallest patio is a 2 x 2.

Facilitation

Guide students to consider what information is important to include in their presentations. Also, encourage the inclusion of representations useful to support statements made. Some students may want to present primarily in a narrative style, whereas others may prefer to present solely with tables and rules, etc. If these kinds of differences arise, use them as opportunities to discuss advantages and disadvantages of each type of representation.

If you plan to use *The Patio Project* as a performance assessment, inform students which criteria you plan to assess. *What to Look For in Lesson 12*, p. 162, provides ideas.

Making the Lesson Easier

Share results after students build three or four patios. Discuss how to predict the numbers of tiles needed for the next-size patio. Refer to the diagrams or models, and use words, not equations, to describe the patterns.

Making the Lesson Harder

Move students toward an equation that represents the total number of tiles for any size patio.

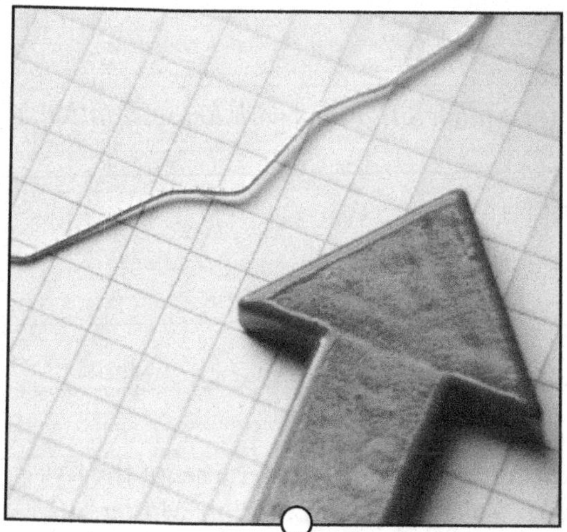

Facilitating Closing the Unit: Putting It All Together

Where do I go from here?

Synopsis

The *Final Assessment* is an opportunity for students to demonstrate what they have learned about tables, graphs, rules, and equations. Anticipate needing two class sessions if you choose to complete all three assessment tasks and *Mind Map and Unit Reflection*. In addition to assessing written work, you will be making observations during this session.

1. Individuals represent a personal pattern with a table, graph, and equation.
2. Individuals match five situations with their corresponding representations and solve related problems.
3. Individuals complete *Symbol Sense Review*.
4. Individuals re-visit the Mind Map of algebra and offer comments on what they have learned in the unit.
5. The class shares thoughts about the unit and next steps in their study of mathematics.

Heads Up!

Though you will assign assessment tasks to individuals, expect that students will share ideas, as they have throughout the unit. Notice those who rely heavily on others to complete work, and bear this in mind when writing up your summative evaluations.

Objective

- Use a variety of algebraic representations to understand situations and solve problems

Materials/Prep

- Colored markers
- Large envelopes (one per student for *Match Situations with Representations*)
- Graph paper
- Rulers
- Scissors
- Tape or staplers

Make sufficient copies of the *Final Assessment*, pp. 191–200, so each student receives a full set of tasks.

For the *Match Situations with Representations* task, clip together a set of the five situations for each student (do *not* staple). Make copies of *Blackline Masters 10* and *11*, and cut up the seven tables, five graphs, and seven equations that students will match to the five situations; put them into a large envelope.

Make sufficient copies of the *Final Assessment Checklist*, p. 201, to accompany each student's completed work.

Note: The personal patterns listed for students in *Task 1*, p.191, represent work from *Lesson 1* done by students in a Massachusetts adult education center and supplied by the teacher. Because classmates' names and habits provided the context, the students had fun. Rather than use the Massachusetts students' list, retrieve appropriate patterns your class described in *Lesson 1*.

Opening Discussion

To prepare for the assessment tasks, ask everyone to turn to p. 151 in the *Student Book*. Read the introduction together, and remind the class that they will work on the tasks in the order outlined:

- *Some Personal Patterns*
- *Match Situations with Representations*
- *Symbol Sense Review*

Encourage students to work independently.

Activity 1: Unit Assessment

Task 1: Some Personal Patterns

Start the assessment tasks by reviewing the directions for *Task 1: Some Personal Patterns*, p. 191. Ask students to choose one of the six personal patterns shared by students in another algebra class. Or, preferably, substitute the patterns from your class's *Lesson 1*. Indicate that there is extra graph paper available for anyone who needs it.

Remind students to be as thorough as possible with their graphs and as clear as possible when writing their rules and equations.

Task 2: Match Situations with Representations

As students complete *Some Personal Patterns*, start them on *Task 2: Match Situations with Representations* and *Solve the Problems*. They attach the matching tables, equations, and graphs you provide in envelopes to the appropriate situation pages, and then solve the problems.

During this task, pay close attention to the processes students use in matching representations because these reveal more than the final products. Ask:

How do you know this table (or graph, or equation) relates to a particular situation?

Specifically, focus on the situations that involve multiple patterns—the race and the fund-raising scenarios. Do not, however, attempt to correct decisions that students have made. Save your comments for later.

Remind everyone to mark the situations "linear" or "nonlinear" and to indicate the constant rate of change for all linear situations. Ask:

What information guides your decisions about the constant rate of change?

Is there another way you can see the rate of change?

As students work on solving the problems, encourage them to show all work, so you can better understand their thought processes.

Task 3: Symbol Sense Review

This task is straightforward.

Activity 2: Mind Map and Unit Reflection

Leave 15–20 minutes for students to work on the *Mind Map and Unit Reflection* activity (*Student Book*, pp. 152–153). Begin by asking who remembers making a Mind Map on the first day of the algebra unit. Choose a volunteer to explain the mapping process. If no one recalls the process, provide a summary and an example of a Mind Map.

For the *Reflection* piece, encourage students to use drawings as well as words to express themselves.

Summary Discussion

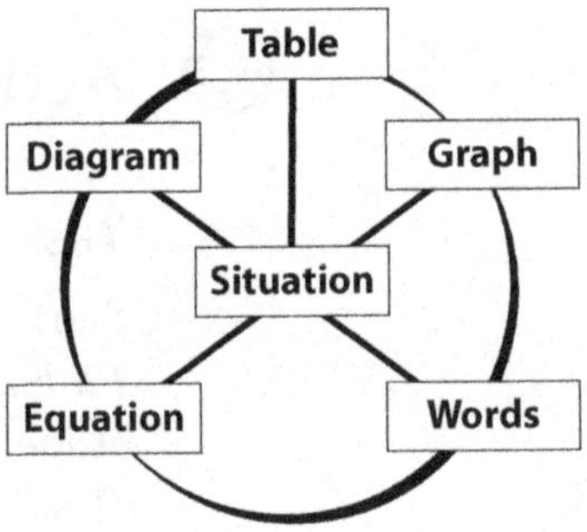

Collect all work other than the *Mind Map and Unit Reflection*. Review any tasks students wish to discuss. This review might include justifications for matching situations with representations, comparing graphs for similar personal patterns, or examining of *Symbol Sense Review* responses. Keep this brief. You will offer individual feedback later. You might also ask:

How did you know which situations were linear or nonlinear, or which ones had the fastest rate of change?

Share the *Mind Map and Unit Reflection* work as a class. If no one offers to share his or her whole Mind Map, draw a map on the board or project a slide with one, and ask volunteers to write on it ideas they linked with the word "algebra."

Students will have written their opinions regarding the most important things they learned about algebra. Bring closure to the unit by inviting everyone to share their reflections. Ask:

What will you remember about algebra?

What would you like to know more about?

Tell the class you will check their work and return it with comments. Urge everyone to continue to look for patterns in the world around them.

Looking Closely

Observe whether students are able to

Use a variety of algebraic representations to understand situations and solve problems

Some Personal Patterns: Here your assessment focuses on the students' ability to generate accurate representations from a given situation. Can students create conventional and accurate graphs? Are they able to set up a table that is useful whether or not it follows convention? Can students apply their new knowledge about rate of change to a particular situation? You also want to note whether students retain the ability to discriminate between looking for the relationship between x and y and finding the rate of change.

Match Situations with Representations: The focus here is interpreting algebraic representations. You want to see whether students connect the various forms accurately, but you also want to observe *how* they do this. Are students able to say *why* a particular equation/table/graph matches the situation? Or are students guessing and using the process of elimination to make their decisions? Notice whether students see that steeper graphs match equations with higher coefficients because the rate of change is faster. How do students work at making connections—do they always match graphs first, or tables, or equations? Why do they make these choices? *All of this information must be gathered as you circulate and observe*. The completed exercise will tell you only whether the student correctly matched the situation with its representations.

Having students identify situations as linear or nonlinear and indicate rates of change for linear relationships provides an opportunity to ask questions about the different clues in the representations that relate to the rate of change. It is important to understand how students make their decisions. Do they pay attention to the graphs, differences in the tables, coefficients in the equations, or all of these?

Solve the Problems: Consider how students solve these problems. The students who show the highest content mastery might be expected to make greater use of equations and might also recognize that some solutions can be found using a variety of representations. Probe to see who has this flexibility to think of multiple ways to write the equations.

Symbol Sense Review: Students at this point should be much more comfortable using notation. Notice whether people interpret notation in a way that facilitates evaluation of expressions. In addition to checking this explicit work with symbols, note how flexible and fluent students are working with notation when problem solving.

Mind Map and Unit Reflection: Compare students' original Mind Maps with those from this activity. Your analysis can be both quantitative and qualitative. You might compare the number of words/ideas that students link with algebra each time to understand how their conceptions of algebra have broadened. By looking at the kinds of words students use and how they link them, you can determine whether their understanding has become more mathematical, more specific, and/or more connected.

Rationale

Although you have engaged in formative assessments throughout the unit, it is important to perform a summative assessment that provides a more complete picture of student achievement. As with all assessments, these exercises are meant to inform instructional decisions. You should have a clear idea of what students do and do not know so you can decide where to focus further practice.

Context

The *Final Assessment* does not address every aspect of all lessons, but focuses heavily on the lessons following the *Interim Assessment*. It provides an opportunity for students to pull together what they have learned about graphs of multiple situations, rate of change, and linear vs. nonlinear situations. However, in *Some Personal Patterns*, students do have a chance to review earlier work about representing situations in various ways.

Facilitation

It is important to observe without instructing. Or if you do instruct, note on the student's work the need to review earlier lessons.

The choice to facilitate responses for all or some of the activities is yours. You may want to simply collect and review student work, or you may feel students need or want to discuss answers. Debriefing the activities can require much time. Choose to share strategies and responses for those activities you feel would be most helpful to the general class.

Summarizing is, however, crucial here. Students are ending an entire unit. You want to know how they are thinking about algebra at this point (*Mind Map*). You also want to allow time for students to share what they have learned from the lessons and how they might apply it in their everyday lives. Hearing what they want to know more about may further guide your instructional decision-making.

Making the Lesson Easier

Pair students.

Comparing student responses provides a general picture of classroom learning as well as insights into particular areas of strength and weakness for individuals.

Some Personal Patterns

You will see cases where a student shows some understanding of the pattern and ways it can be represented, but also areas that need attention. However, some students progress to work equally well with all representations.

Case A

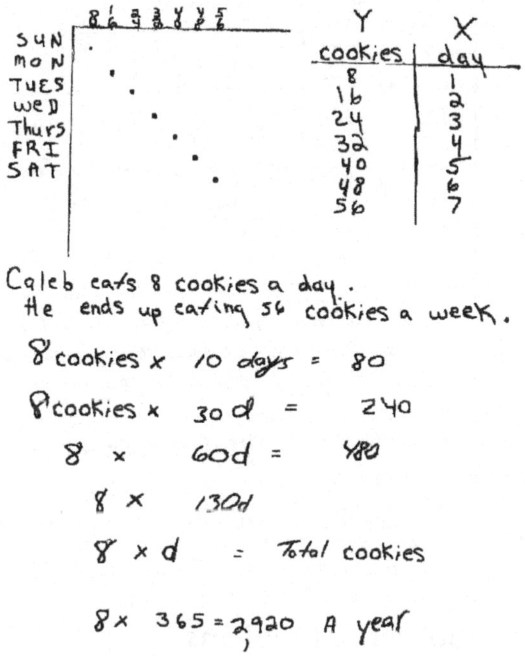

It seems picky to point out the graph's origin in the upper-left corner and the mathematically unnecessary use of "d" in this student's equations, given the strong connection between his four highly communicative representations. On the other hand, we acknowledge that adults need to know standard conventions.

Holyoke Adult Learning Organization (HALO)
Holyoke, MA

continued on next page

continued from previous page

Case B

Table.

x no. of cookies	y no. of C. eaten daily
1	8
2	16
3	24
4	32
5	40
6	48
7	56
8	64
9	72
10	80

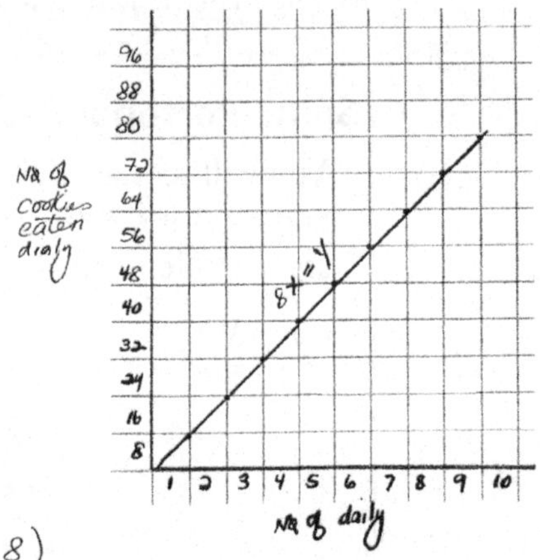

Equation = 8x = y (1x8 = 8)

Rule = 8 cookies represents one day of the week (multiply no. of days by cookies).

3 days = 24
5 〃 = 40
10 〃 = 80.

A rule that you can use to predict what would happen after any number of days. x represents the number of day Caleb eats 8 cookies. y represents the amount of cookies eaten as the days increase.

Rule as an equation = 8x = y

In Case B, the student shows proficiency with all standard representations.

Borough of Manhattan Community College
New York City, NY

Solve the Problems

Problem 3: In the race between James and Shana, who will win?

Case A

6x20+80 = 200

James

x S	M y
1	86
2	92
3	98
4	104
5	110
6	[illegible]
7	118
8	124
9	13[illegible]
10	136

10x20 = 200

Shana

x S	M y
1	10
2	20
3	30
4	40
5	50
6	60
7	[illegible]
8	80
9	90
10	100

The Case A student demonstrates a grasp of the table as a means to solve the problem; however, she does not extend the table far enough to see the solution. The teacher indicated that when the class reviewed the problem, the student wrote in the equations that others had used.

Borough of Manhattan Community College
New York City, NY

continued on next page

continued from previous page

Problem 1: How much money will Franz have for the community center after receiving his 100th donation?

Case B

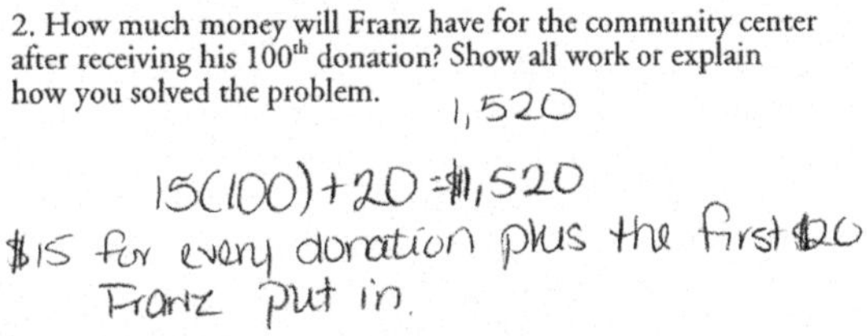
2. How much money will Franz have for the community center after receiving his 100th donation? Show all work or explain how you solved the problem.

1,520

15(100) + 20 = $1,520

$15 for every donation plus the first $20 Franz put in.

Because the student in Case B is comfortable with equations and understands that the 100th case is the x-value in this case, she easily solves the problem.

Borough of Manhattan Community College
New York City, NY

Mind Maps

Because of open enrollment, algebra Mind Maps can vary widely from student to student. Below are two examples of Mind Maps. We notice that the lessons spark particular details and perspectives, but the exercise allows students to tap into prior knowledge. What we also notice is the paucity of linkages. Students connect everything equally to the main idea of "algebra." This may indicate a need for literacy, as well as numeracy, development.

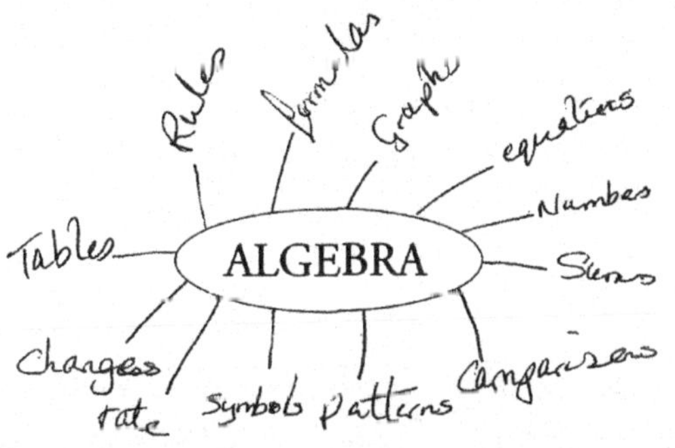

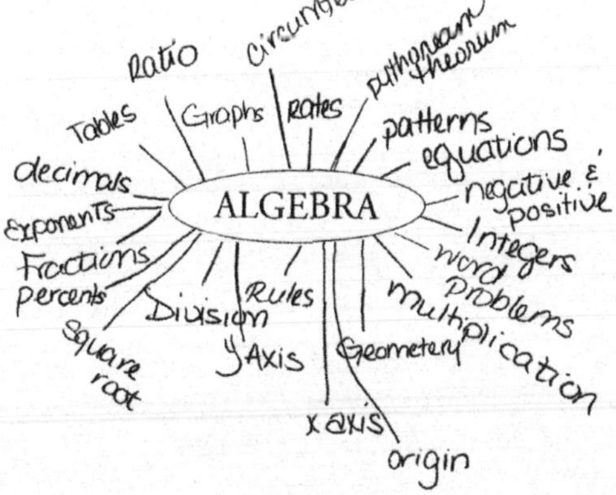

Borough of Manhattan Community College
New York City, NY

Name ______________________________ Date ____________

Seeking Patterns, Building Rules: Algebraic Thinking

INITIAL ASSESSMENT: PATTERNS AND PROBLEMS

Look at each task and decide whether you can do it, do not know how to do it, or are not sure whether you can do it. Check off the appropriate choice, then complete the task if you can.

Task 1: Visual Patterns

1. Think about what is happening in this pattern.

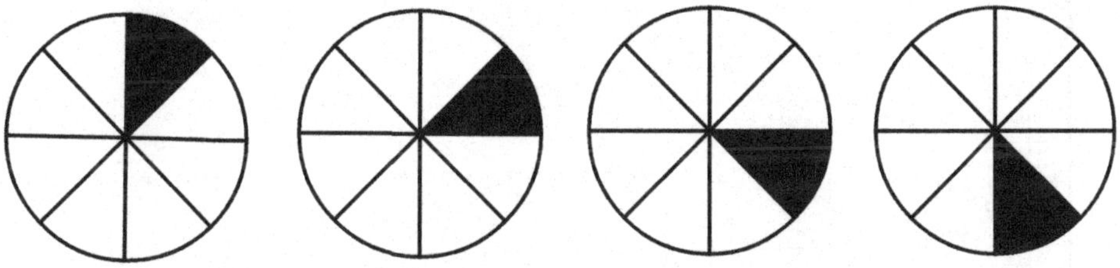

   a. What will the next picture look like? Draw it.

   ___ Can do ___ Don't know how ___ Not sure

   b. What will the 50th picture look like? How do you know?

   ___ Can do ___ Don't know how ___ Not sure

2. Which sentence best describes this pattern ?

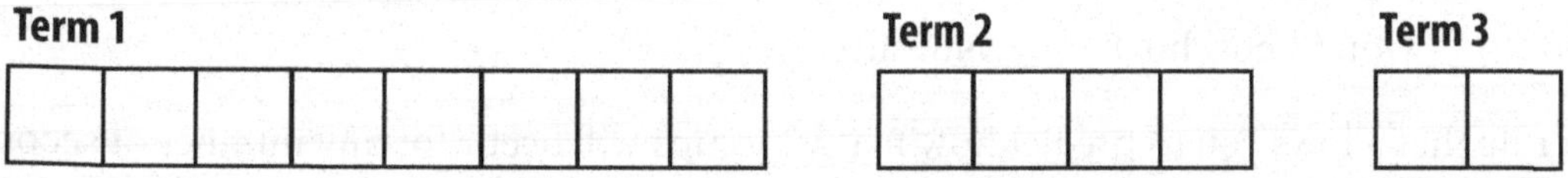

   a. As the term number increases by one, the number of squares increases by two.
   b. As the term number increases by one, the number of squares is reduced by two.
   c. As the term number increases by one, the number of squares is reduced by one.
   d. As the term number increases by one, the number of squares increases by four.
   e. As the term number increases by one, the number of squares is divided by two.

   ___ Can do ___ Don't know how ___ Not sure

Task 2: Number Patterns

Fill in the missing numbers in the table.

x	*y*
2	11
3	14
5	
	35

___ Can do ___ Don't know how ___ Not sure

Task 3: Make a Graph and a Rule

Hiccups

Seconds	Hiccups
2	1
4	2
6	3

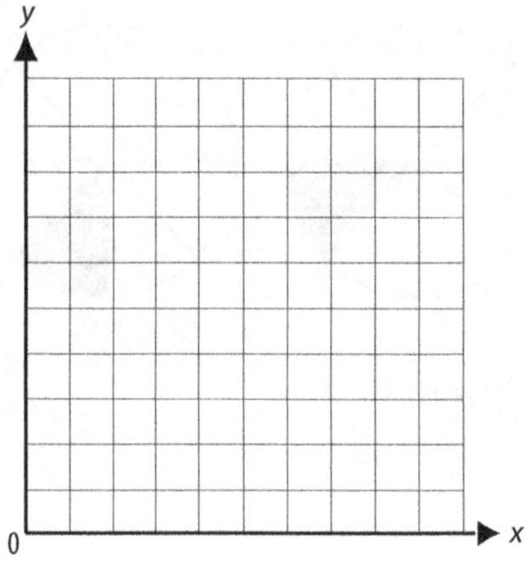

1. Complete the "Hiccups" table by making another entry in the "Seconds" and in the "Hiccups" columns.

___ Can do ___ Don't know how ___ Not sure

2. Use the table data to make a graph.

___ Can do ___ Don't know how ___ Not sure

3. Write a rule that allows you to predict how many hiccups will occur for any number of seconds.

___ Can do ___ Don't know how ___ Not sure

4. Write an equation that represents your rule. Let *x* stand for the number of seconds. Let *y* stand for the number of hiccups.

___ Can do ___ Don't know how ___ Not sure

Task 4: Matching Graphs, Tables, and Equations

Match the graphs with the appropriate tables and equations.

___ Can do ___ Don't know how ___ Not sure

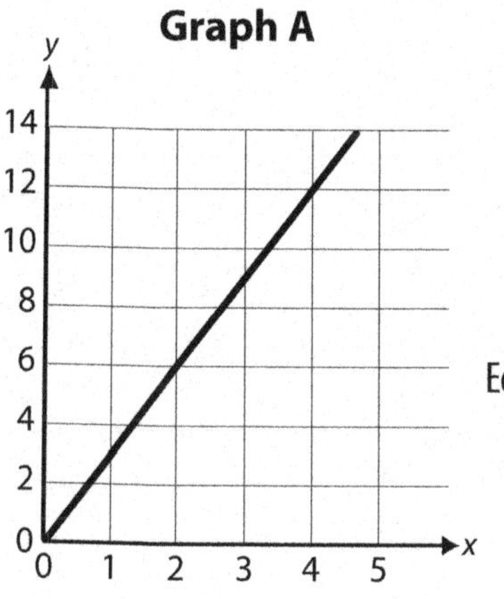

Table ______

Equation ______

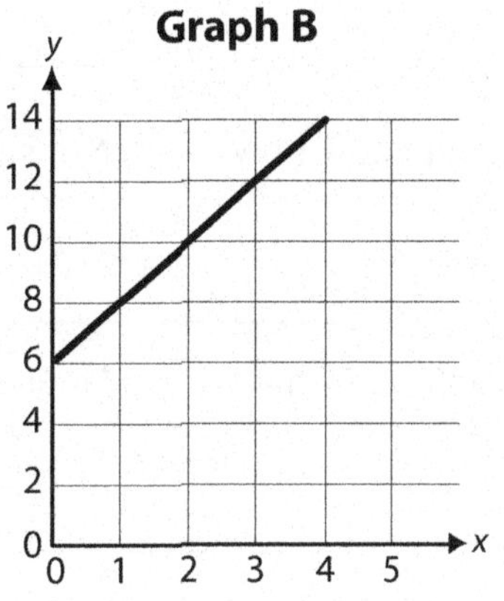

Table ______

Equation ______

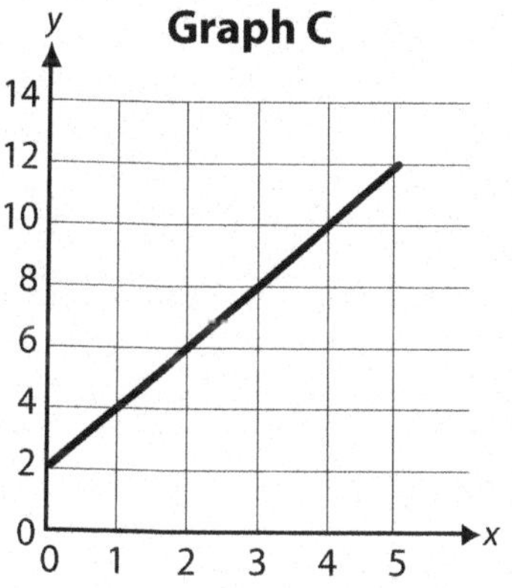

Table ______

Equation ______

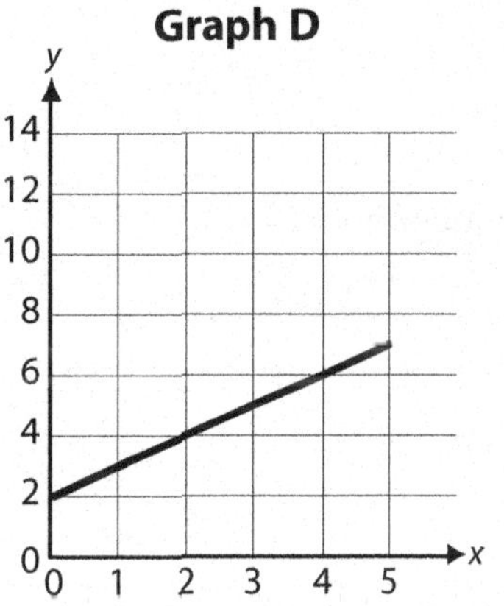

Table ______

Equation ______

Table 1

x	y
1	8
2	10
4	14

Table 2

x	y
1	3
2	6
3	9

Table 3

x	y
1	3
2	4
4	6

Table 4

x	y
1	4
2	6
5	12

Equation 1: $y = 3x$

Equation 2: $y = x + 2$

Equation 3: $y = 2x + 2$

Equation 4: $y = 2(x + 3)$

Task 5: Symbol Sense

1. Mark each statement "True" or "False." If the statement is false, rewrite it as a true statement.

a. $3(15) = 3 + 15$ ________________________

___ Can do ___ Don't know how ___ Not sure

b. $\frac{40}{10} = \frac{10}{40}$ ________________________

___ Can do ___ Don't know how ___ Not sure

c. $5 + 4(10) = 90$ ________________________

___ Can do ___ Don't know how ___ Not sure

d. $5(4) + 20 = 40$ ________________________

___ Can do ___ Don't know how ___ Not sure

e. $8^2 = 8 \times 8$ ________________________

2. Solve for x in each equation:

a. $x + 12 = 42$ $x =$ ________

___ Can do ___ Don't know how ___ Not sure

b. $x - 12 = 42$ $x =$ ________

___ Can do ___ Don't know how ___ Not sure

c. $6x = 42$ $x =$ ________

___ Can do ___ Don't know how ___ Not sure

d. $42 = 5x + 12$ $x =$ ________

___ Can do ___ Don't know how ___ Not sure

INITIAL ASSESSMENT CLASS TALLY

Task	Can Do	Don't Know How	Not Sure	Percent Who Can Do
1. 1a.				
1b.				
2				
2.				
3. 1.				
2.				
3.				
4.				
4.				
5. 1a.				
1b.				
1c.				
1d.				
1e.				
2a.				
2b.				
2c.				
2d.				

Notes on Confidence Levels

INITIAL ASSESSMENT CHECKLIST

Use a ✓, ✓+, or ✓– to assess how well students met each skill. When you give feedback to students, note areas in which they did well in addition to areas for improvement.

Use ✓ to show work that is mostly accurate; some details, additional work needed

Use ✓+ to show work that is accurate, complete.

Use ✓– to show work that is inaccurate, incomplete.

Student's Name____

Task	Skills	Lesson Taught
1. Visual Patterns		
1a. ____	Extends a visual pattern	2
1b. ____	Extends a visual pattern	
2. ____	Describes and generalizes the pattern	
2. Number Patterns	Completes In-Out Table	1, 2, 3
3. Make a Graph and a Rule		
1. ____	Completes In-Out Table	1, 2, 3
2. ____	Makes a table from graph	4–12
3. ____	Writes rule in words from table	All
4. ____	Writes equations for the table	All
4. Matching Graphs, Tables, and Equations	Matches the graphs, tables, and equations	4–12
5. Symbol Sense		
1a. ____	Identifies a(b) as multiplication	1
1b. ____	Knows a/b is not the same as b/a	1
1c. ____	Evaluates a + b(c)	2
1d. ____	Evaluates a(b) + c	2
1e. ____	Solves for *x* in one-step equations (addition)	8
2a. ____	Solves for *x* in one-step equations (subtraction)	8
2b. ____	Solves for *x* in one-step equations (multiplication)	8
2c. ____	Solves for *x* in one-step equations (multiplication and addition)	
2d. ____		9

OVERALL NOTES

Strengths

Areas for Improvement

Name ______________________________ Date ______________

Seeking Patterns, Building Rules: Algebraic Thinking

INTERIM ASSESSMENT: USING THE TOOLS OF ALGEBRA

Task 1: Working from a Graph

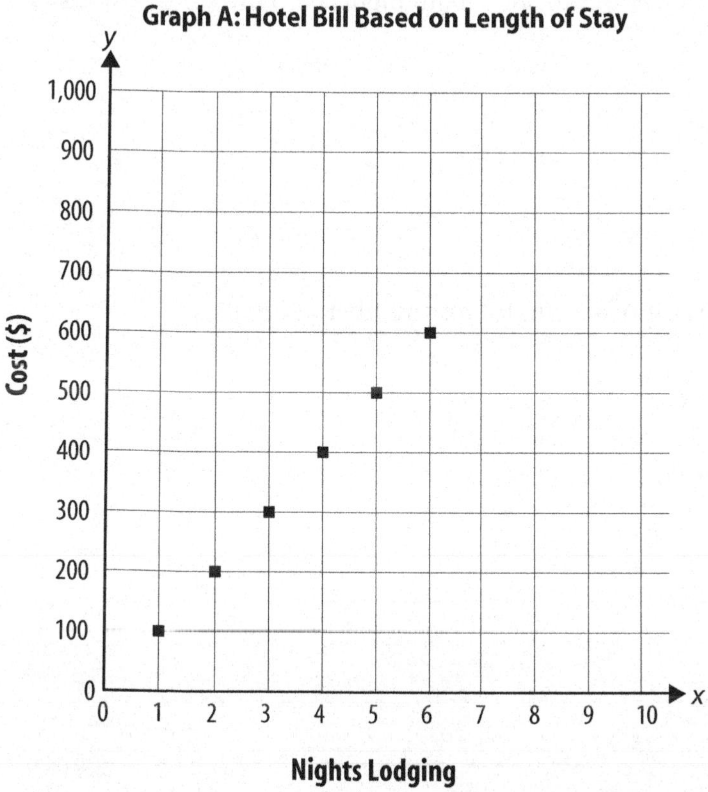

1. Examine Graph A. What does it tell you? Write a brief paragraph that explains the story behind the graph, what situation it describes, and the details of that situation.

2. Make a table from the graph data points.

3. What do you predict would come next on your table?

4. What would the total accumulated cost for the 10th, 20th, and 100th nights be? How do you know?

5. Write a rule that explains how to predict the cost of a room for *any* number of nights.

a In words:

b. As an equation:

Task 2: Working from a Table

Effects of Drought on Reservoir Water Level (Depth)

Duration of Drought (in weeks)	Depth of the Reservoir (in feet)
0	20
1	19
2	18
3	17

1. Examine the table above. What does it tell you? Write a brief paragraph that explains the story behind the table, the situation it describes, and the details of that situation. Be creative, but be accurate too.

2. Make a graph from the table data.

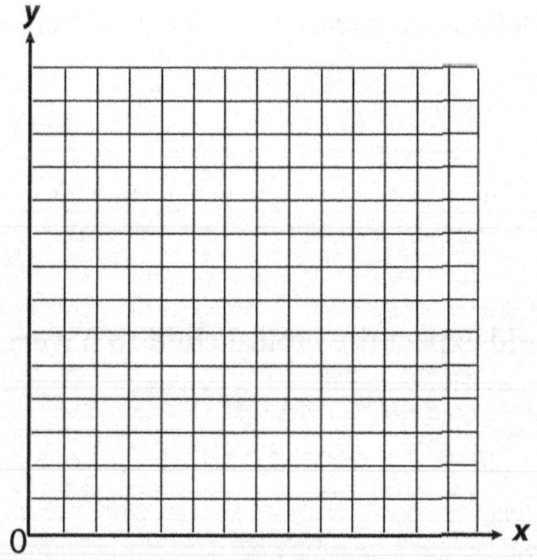

3. What will the reservoir depth be on the 10th week and the 15th week? Show all work.

4. When will the reservoir run out of water? How do you know?

5. Write a rule that explains how to predict the reservoir's depth in feet after *any* number of weeks of drought.

   a. In words:

   b. As an equation:

Task 3: Working from an Equation

The formula for determining the vocabulary size of a typical child between the ages of 20 months and 50 months is

$$y = 60x - 900$$

y = the number of words

x = age in months

1. At the age of 30 months, you could expect the typical child to have a vocabulary of how many words? Show how you know.

   a. 600

   b. 800

   c. 900

   d. 1,800

   e. 2,700

2. Make an In-Out table showing the ages of some children from 20 to 50 months old, and the corresponding vocabulary sizes of the children.

3. Use the vocabulary formula to plot three points on the graph below. Label and mark your x- and y-axes with appropriate scale numbers.

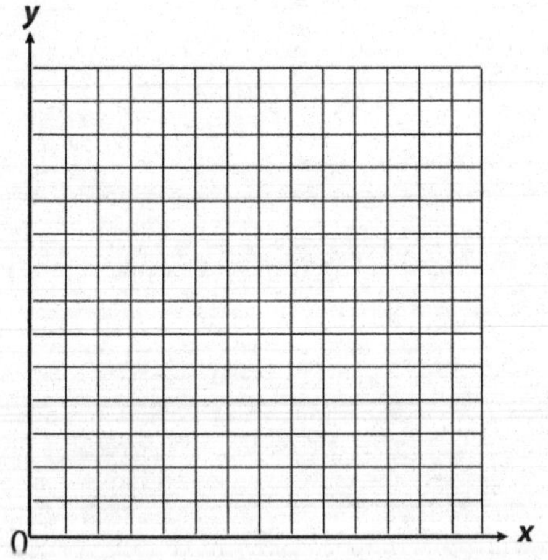

Task 4: Tina's Walking Rate

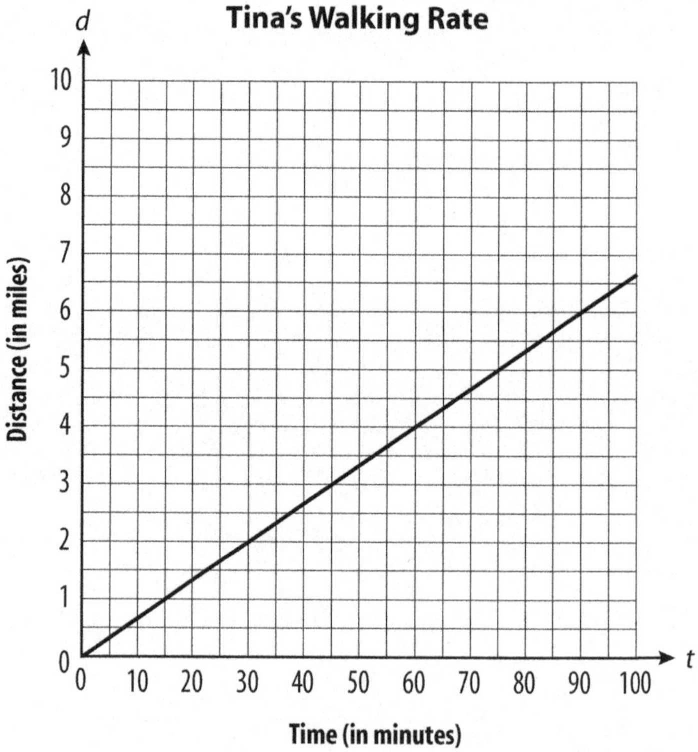

1. Given the information on the graph above, which of the following equations allows you to predict how far Tina will walk in any number of minutes. Let t stand for the time in minutes she walks; let d stand for the distance in miles she walks in that time. Show all work.

a. $d = t + 15$

b. $d = t - 15$

c. $d = \frac{t}{15}$

d. $d = 15t$

e. $td = 15$

Task 5: Symbol Sense

Examine each pair of mathematical statements. Only one is correct. Cross out the incorrect statement and rewrite it correctly. The first one is done as an example.

1. $450 = 45(10)$ 　　　　　　　　　　　　　450/10 = 45

   ~~450(10) = 45~~

2. $3g = 3 \times g$ ____________________

   $20m = 20 + m$

3. $22 - 4(5) = 90$ ____________________

   $30 + 6(3) = 48$

4. $10(5) - 10(4 + 1) = 0$ ____________________

   $10(6 + 1) = 17$

5. $6 + 10 > 7 + 10$ ____________________

   $8(2) < 7 + 10$

6. If $x = 7$, then $100x + 50x = 1{,}050$ ____________________

   If $x = 7$, then $50 + 100x = 1{,}050$

Rewrite:

7. $900 - 200 = 700$ as an addition equation. ____________________
8. $ab = 30$ as a division equation. ____________________
9. $4x + 2 = 16$, in words. ____________________
10. y is equal to six added to x, as an equation. ____________________

Name ______________________________ Date ______________

Seeking Patterns, Building Rules: Algebraic Thinking

FINAL ASSESSMENT

Task 1: Some Personal Patterns

Students in one EMPower algebra class shared the following personal patterns …

- Emily attends the Adult Learning Center five times a week.
- Monday through Friday, Siomara works from 9:30 a.m. to 5:30 p.m., 40 hours per week.
- Jason eats two bowls of corn pops a day.
- Evelina drinks three cups of coffee a day.
- Miriam's goddaughter drinks four ounces of milk in the morning and six ounces at night.
- Caleb drinks eight glasses of water a day.

1. Choose one situation to work with for this exercise.
2. On the next two pages, make a table, a graph, a rule in your own words, and an equation for the situation.
   - Make a *table* to show what the person uses or does over a period of time, such as over many days or weeks.

     ___ Can do ___ Don't know how ___ Not sure
   - Make a *graph* that represents your table data.

     ___ Can do ___ Don't know how ___ Not sure
   - Write a *rule* you could use to predict what would happen after any number of days (or weeks).

     ___ Can do ___ Don't know how ___ Not sure
   - Write the rule as an *equation.* Express the equation two different ways.

     ___ Can do ___ Don't know how ___ Not sure
   - Show the rate of change on your graph. Describe where you see the rate of change in the table and in the equation.

     ___ Can do ___ Don't know how ___ Not sure

The situation I chose:

My table:

The rule in my own words:

My equation:

My Graph

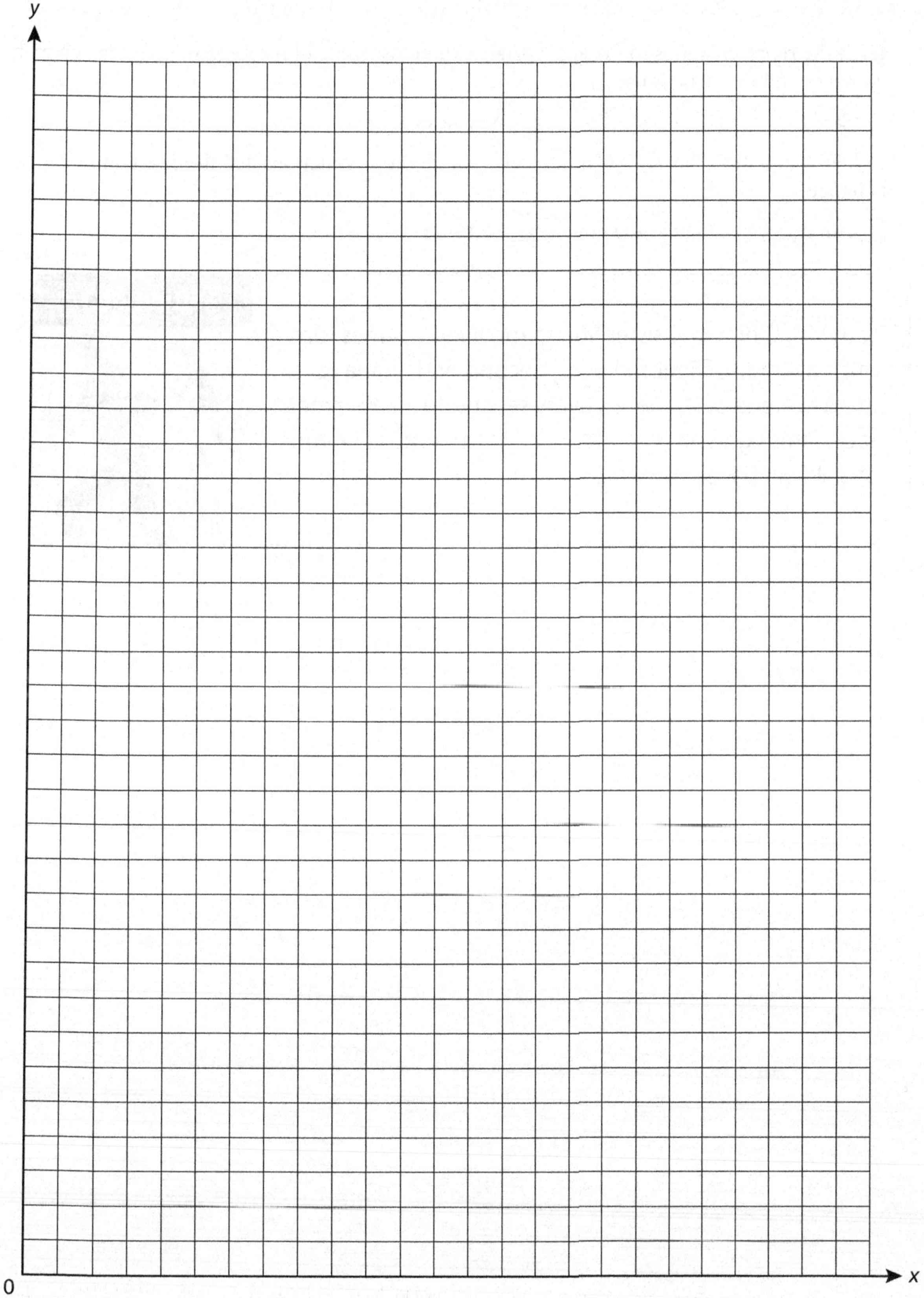

Task 2: Match Situations with Representations

Following are five situations. In the folder you will find five graphs, seven tables, and seven equations.

- Match the representations to the appropriate situations. Some of the situations will match with two tables and two equations.

  ___ Can do ___ Don't know how ___ Not sure

- Mark each situation "linear" or "nonlinear." For all linear situations, tell the constant rate of change.

  ___ Can do ___ Don't know how ___ Not sure

Situation 1

James and his older sister, Shana, run a race. James runs at an average of 6 meters every second, and Shana runs at an average of 10 meters every second. In a 200-meter race, James gets an 80-meter head start because he runs at a slower pace.

Situation 2

Sam shovels snow for his apartment building, charging $40 per snowfall. The first time he shoveled, there was a blizzard, and it took him three hours. The second time the snow was light, and it took him a half-hour.

Situation 3

At Tile World, the price of a tile depends on its surface area. A 1-inch by 1-inch tile sells for $0.10; a 2-inch by 2-inch tile sells for $0.40; and a 3-inch by 3-inch tile sells for $0.90, and so on.

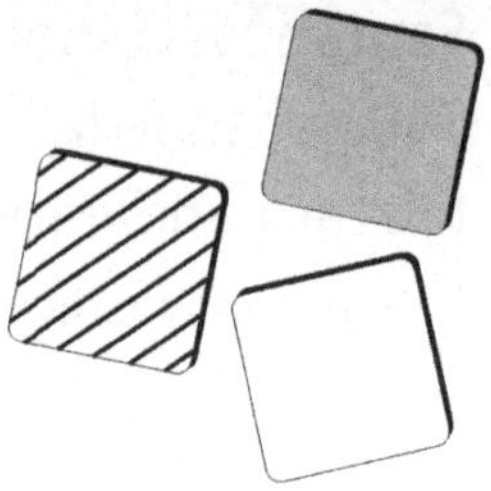

Situation 4

Alphonso works six hours each night of the week, weekends included, at his part-time job. This way, he can stay home days to care for his children. He keeps track of all the hours he has worked since he started.

Situation 5

Tanya and Franz are trying to raise funds for a new community center. Tanya asks every person she meets to donate $20. Franz put in $20 of his own money and is asking each person he meets to donate $15.

Solve the Problems

Now that you have completed the matching portion of this activity, it is time to solve some problems related to the situations above.

> Use any and all of the representations to help you solve the problems. Extend the graphs or tables if that is helpful.

1. How much money will Franz have for the community center after receiving his 100th donation? Show all work, or explain how you solved the problem.

   ___ Can do ___ Don't know how ___ Not sure

2. Franz and Tanya have both collected $80 after four donations. When will they next show the same number of dollars collected? Have they received the same number of donations? Show all work, or explain how you solved the problem.

   ___ Can do ___ Don't know how ___ Not sure

3. In the race between James and Shana, who will win? Show all work, or explain how you solved the problem.

   ___ Can do ___ Don't know how ___ Not sure

Task 3: Symbol Sense Review

1. Evaluate each expression. Let $n = 12$, $r = 8$.

   a. $6 + 3n$

   ___ Can do ___ Don't know how ___ Not sure

   b. $5nr$

   ___ Can do ___ Don't know how ___ Not sure

   c. $2n + 3r$

   ___ Can do ___ Don't know how ___ Not sure

   d. $\frac{n-r}{2}$

   ___ Can do ___ Don't know how ___ Not sure

   e. $(n + r)(n - r)$

   ___ Can do ___ Don't know how ___ Not sure

   f. nr^2

   ___ Can do ___ Don't know how ___ Not sure

   g. $(nr)^2$

   ___ Can do ___ Don't know how ___ Not sure

   h. r^2n^2

   ___ Can do ___ Don't know how ___ Not sure

2. Solve each equation for x:

   a. $3x + 15 = 75$

   ___ Can do ___ Don't know how ___ Not sure

   b. $5x - 8 = 47$

   ___ Can do ___ Don't know how ___ Not sure

   c. $5(x + 9) = 200$

   ___ Can do ___ Don't know how ___ Not sure

FINAL ASSESSMENT CHECKLIST

Use a ✓, ✓+, or ✓– to assess how well students met each skill. When you give feedback to students, note areas in which they did well in addition to areas for improvement.

Use ✓ to show work that is mostly accurate; some details, additional work needed.

Use ✓+ to show work that is accurate, complete.

Use ✓– to show work that is inaccurate, incomplete.

Student's Name______________________________

Task	Skill	Lesson Taught
1. Some Personal Problems	Builds a representative table Makes a complete graph of a personal pattern Describes relationship between variables Expresses relationship between variables as algebraic equations Shows rate of change on graph and describes where it is located in the table and equation	2, 5, 6, 8 4, 8 All All 10–12
2. Match Situations with Representations	Matches situations with representations and offers reasonable explanations for choices Connects representations with linearity and nonlinearity Quantifies constant rates of change	2–12 11, 12 10–12
Solve the Problems	Uses representations to accurately solve problems	2, 3, 6, 8, 9, 12
3. Symbol Sense Review	Understands meaning of notation Evaluates expressions with rules of order Understands inverse relationships Solves equations in one variable	All 2, 3, 7 4, 5 8, 9, 10

OVERALL NOTES

Strengths

____

____

____

____

____

____

____

Areas for Improvement

Blackline Master 1

The Waitress's Problem

Kim was waitressing for the first time. She did not know what to expect for tips. A veteran waitress told her, "Usually folks here give a 25% tip." Kim must have looked confused; the experienced waitress ripped off a piece of paper from her check pad and scribbled on the back:

Bill	Tip
$1	$0.25
$2	$0.50
$3	$0.75

"It just goes like that," she said, as she handed the piece of paper to Kim. Kim thanked her, thinking, "Yeah, but what should I expect to get for a bill that totals $7.00?"

Blackline Master 3

Guess My Rule Tables

Table 1

x	y
2	6
3	9
4	12

Table 2

x	y
3	120
5	200
7	280

Table 3

x	y
56	7
96	12
104	13

Table 4

x	y
1	
18	24
21	27
	40

continued on next page

Blackline Master 3 (continued)

Guess My Rule Tables

Table 5

x	*y*
2	0
3	
4	2
	12

Table 6

x	*y*
1	3
2	5
4	
	17

Table 7

x	*y*
1	4
2	6
4	10
	220

Table 8

x	*y*

Blackline Master 4
Secret Equation Cards

$y = 12x$	$y = \frac{x}{20}$
$y = 7x$	$y = x$ $(y = 1x)$
$y = x - 12$	$y = \frac{x}{7}$
$y = x + 1$	$y = x + 20$
$y = 12x + 25$	$y = 20 - x$

Blackline Master 5

Pass the Message Trifolds

Equation

Rule in Words

Equation

Equation

Rule in Words

Equation

Blackline Master 6

Phone Plans

Cellular *wireless*

7 cents per minute

No Monthly Charge!

Z-phone

250 Anytime Minutes

$15 per month

(30 cents each additional minute)

Tel-Mobile *wireless*

$5.00 Monthly Fee

5 cents/minute

NuWave *Wireless*

1,000 Anytime Minutes for $40.00/Month

$0.35 each additional minute

$C = \$5.00 + \$0.05M$

$C = \$0.07M$

Monthly Minutes	Cost
50	$3.50
100	$7.00
200	$14.00
500	$35.00

Monthly Minutes	Cost
50	$7.50
150	$12.50
250	$17.50
350	$22.50

Monthly Minutes	Cost
100	$15.00
150	$15.00
250	$15.00
300	$30.00

Monthly Minutes	Cost
500	$40.00
1,000	$40.00
1,100	$75.00
1,200	$110.00

for $M \leq 250$, $C = \$15$
for $M > 250$, $C = \$15 + \$0.30(M - 250)$

for $M \leq 1{,}000$, $C = \$40$
for $M > 1{,}000$, $C = \$40 + \$0.35(M - 1{,}000)$

Blackline Master 6a

Phone Plans (alternate version)

Cellular wireless

Unlimited Premium Data
only $70 per month!

One-time activation fee
($35 per device)
Hotspot data not included.

Z-phone

UNLIMITED

Talk, Text & Data only $75 / month!

Guaranteed Hi-Speed Data - No Matter What!

40 GB hotspot data per line per mo.
& premium entertainment pkg
$30 one-time activation fee (per plan)

Tel-Mobile

Unlimited

talk, text, & 4 GB data per line

Only $50/month!*

Incl 4 GB mobile hotspot data/line/mo.

$45 one-time activation fee (per plan)
** Data overage fee: If monthly 4 GB data limit is exceeded, a $10 fee for every 2 GB additional data applies.*

NuWave Wireless

$40/Month!

UNLIMITED talk, text, & data

- ✱ **NO activation fee**
- ✱ **NO upgrade fee**
- ✱ **FREE unlimited mobile hotspot data!**

Hotspot speed capped at 5 Mbps; limit 1 device tethered at a time

---

C = $75M + $30

C = $40M

Number of Months	Total Cost
1	$105
3	$245
6	$455
12	$875

Number of Months	Total Cost
1	$105
5	$405
9	$705
12	$930

Number of Months	Total Cost
1	$95
4	$245
8	$445
12	$645

Number of Months	Total Cost
1	$40
3	$120
6	$240
12	$480

C = $35 + $70M

C = $50M + $45

Activity 1: Phone Plans

Four ads for wireless phone plans came in the mail today, but the ads got ripped up. Can you put the pieces back together? Your teacher will give you all of the words, tables, and equations from the ads. Reattach them to their matching graphs.

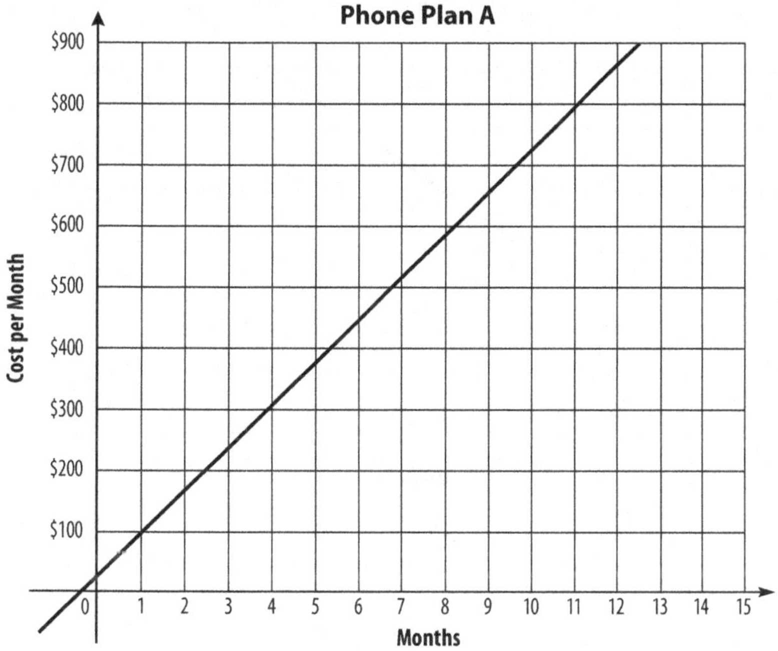

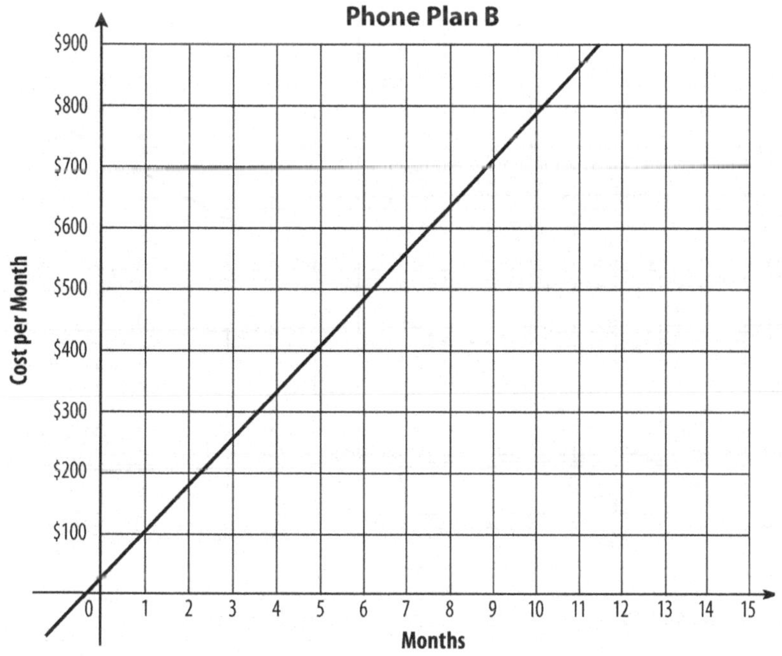

Blackline Master 6a (continued)

Phone Plans (alternate version)

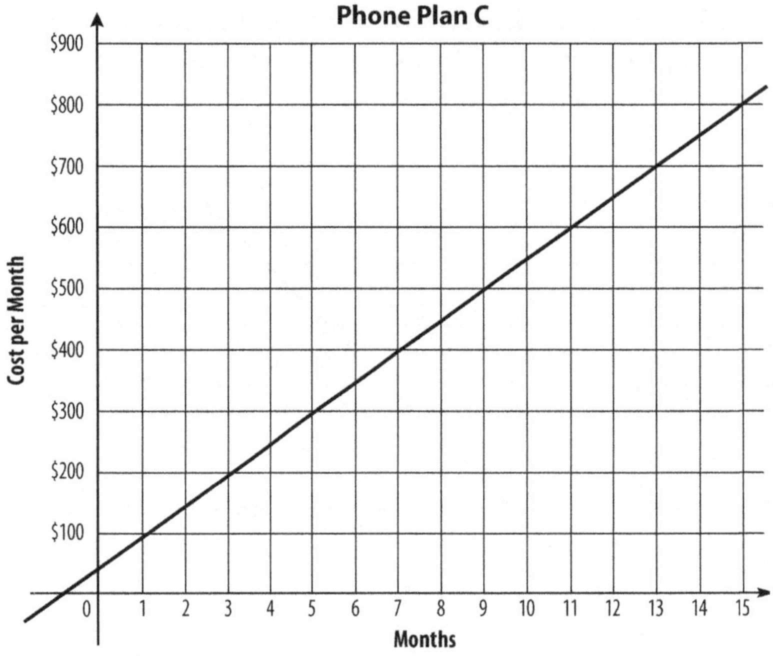

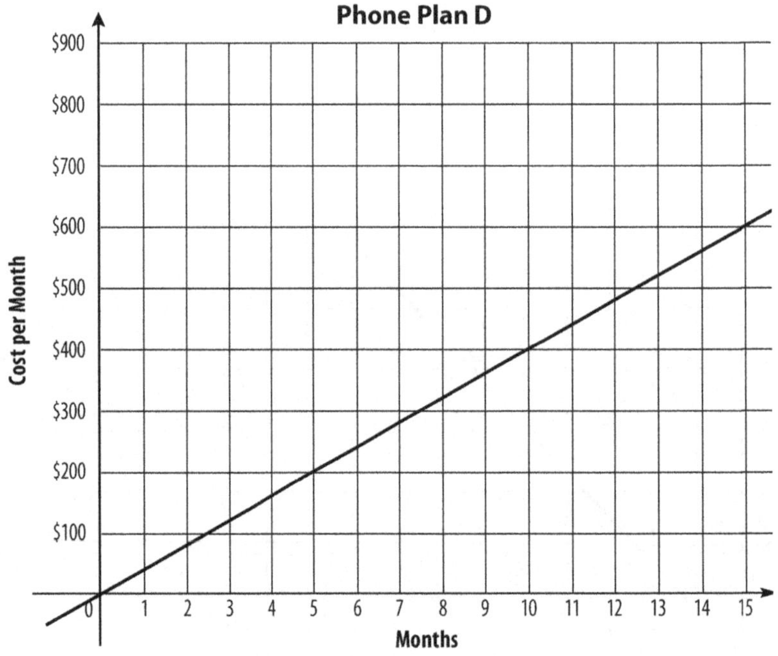

Activity 2: It Would Depend on the Person

Use the pieced-together advertisements to answer the following questions about which plans are best and worst for specific customers. Note: In addition to the information shown on the ads, each plan also allows customers to save more per month with additional lines.

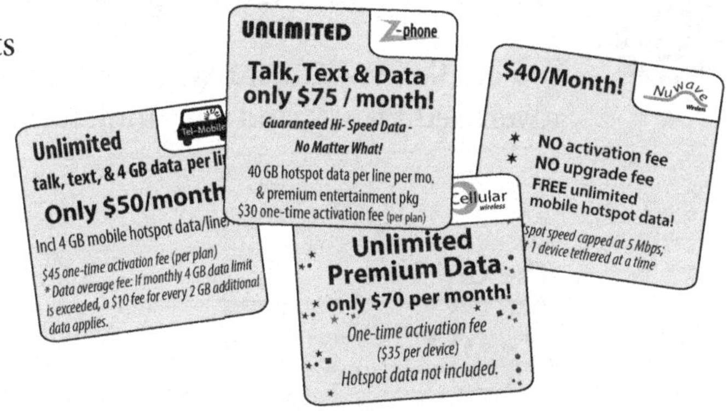

1. Mary Jane loves to talk and text her family, and keep up with the newest viral videos and social media posts. She hates it when it takes forever to view these due to slowed-down data on her phone. She does not want to worry about exceeding her data or having it throttled down. Which plan should she choose? Why? Which plan would be the worst for her? Why?

2. Hassun lives in a rural community where internet service is limited but cell phone service is strong due to a nearby cell tower. He uses his hotspot on his phone to connect his laptop to his online classes. His family also uses the same hotspot for telehealth appointments and videochatting with family and friends using other electronic devices, even while Hassun is in class. Which plan is best for him, and which is the worst? Why?

3. Akari wants to get a phone plan she only will use in emergencies, for instance, if her car breaks down. Which plan is best for her, and which is the worst? Why?

4. Think about your cell phone and data needs and the number of people who might share a phone plan with you. Which plan would be the best for your situation? Which would be the worst? Why? What additional information would you want to know to help you to decide?

Practice: I Am Changing the Rules!

You are the new Chief Executive Officer (CEO) of Cellular Wireless that has advertised Plan A, and you think it is time to change the rules.

The Old Rule

The New Rule

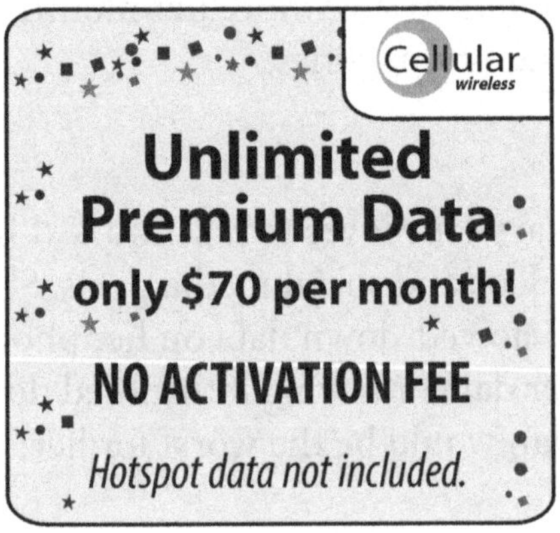

Change the old equation, table, and graph to go along with your new rule for the first month.

The Old Equation

C = $35 + $70M

The New Equation

____

The Old Table

Number of Months	Total Cost
1	$105
3	$245
6	$455
12	$875

The New Table

Blackline Master 6a (continued)
Phone Plans (alternate version)

This is a graph of the old plan.

Graph the new plan in a different color.

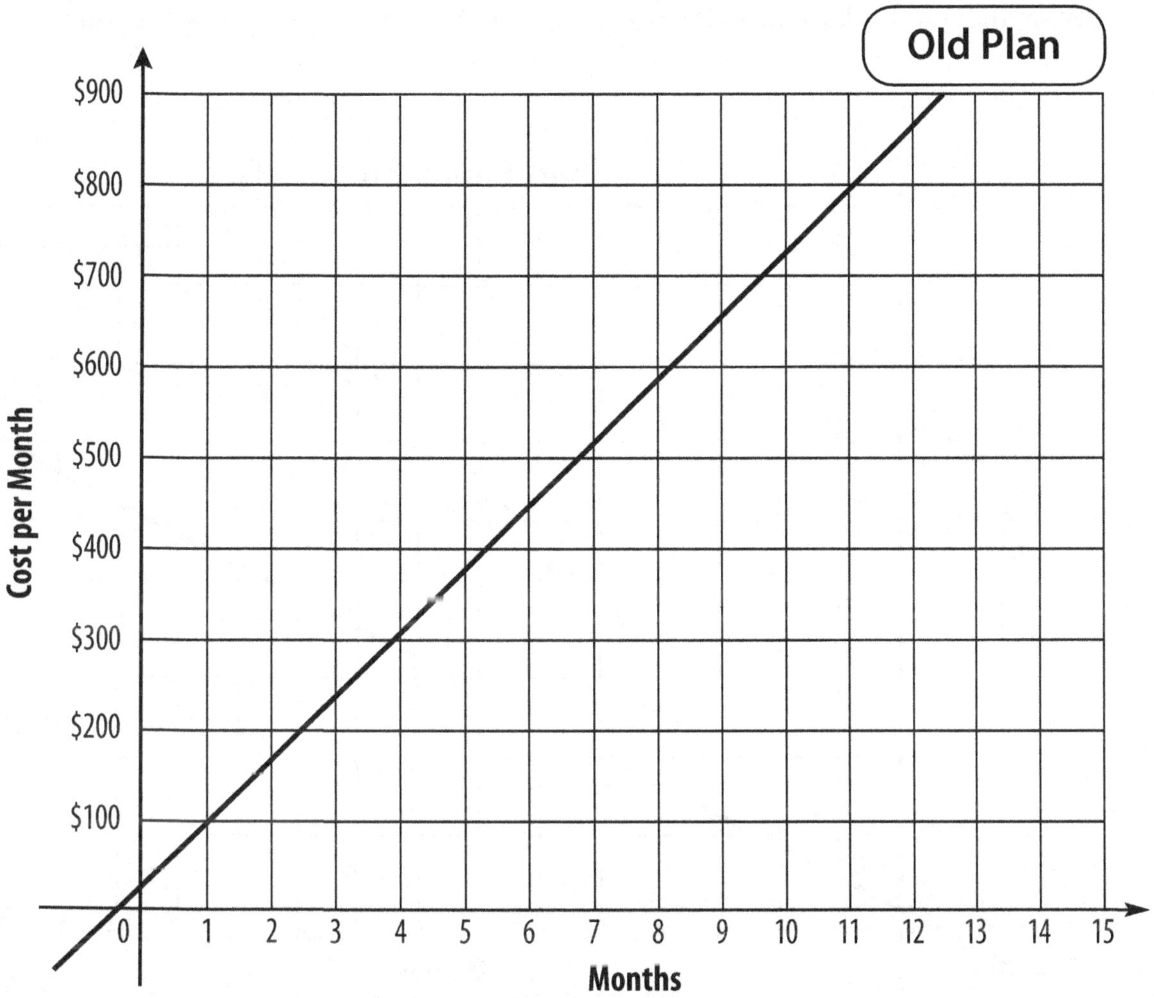

Extension: Looking at Four Graphs

Juania says: "I really do not think it makes much difference for me and the others who will share my plan. On Plans A and B, we will each pay about the same … but Plans C and D are definitely out."

What do you think their phone use pattern is? Use the graph to support your conclusion.

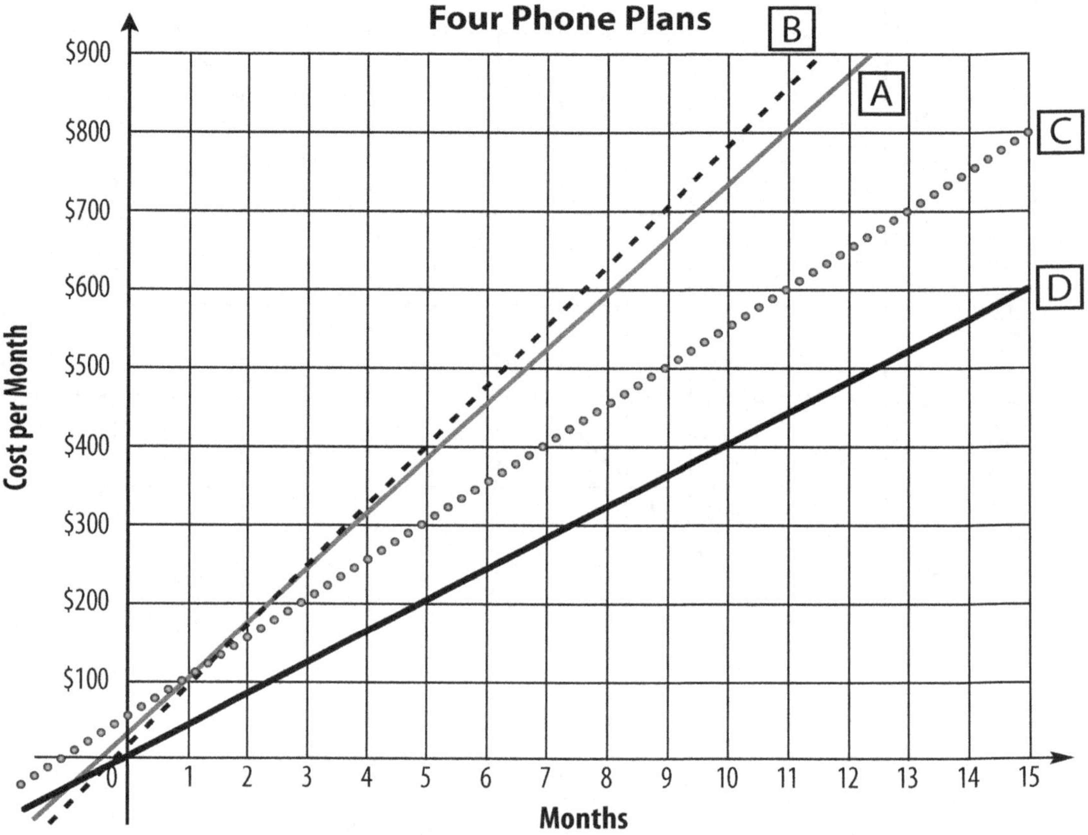

Blackline Master 6a ANSWER KEY

Phone Plans (alternate version)

Lesson 9: Phone Plans

Activity 1: Phone Plans

Phone Plan A

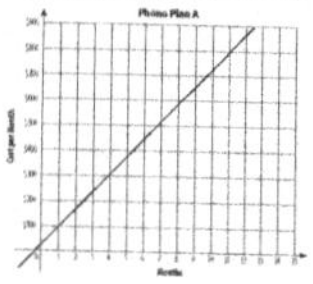

C = $35 + $70M

Cellular wireless

Unlimited Premium Data
only $70 per month!
One-time activation fee ($35 per device)
Hotspot data not included.

Number of Months	Total Cost
1	$105
3	$245
6	$455
12	$875

Phone Plan B

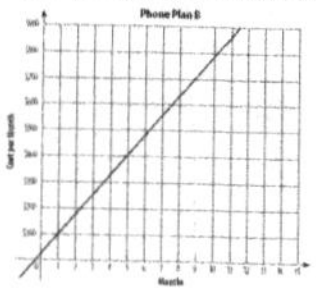

C = $75M + $30

UNLIMITED Z-phone

Talk, Text & Data only $75 / month!
Guaranteed Hi-Speed Data - No Matter What!
40 GB hotspot data per line per mo. & premium entertainment pkg
$30 one-time activation fee (per plan)

Number of Months	Total Cost
1	$105
5	$405
9	$705
12	$930

Phone Plan C

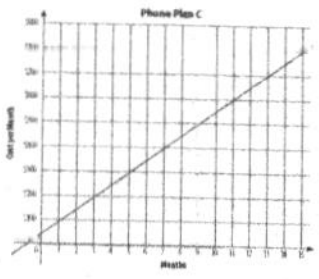

C = $50M + $45

Unlimited Tel-Mobile
talk, text, & 4 GB data per line
Only $50/month!*
Incl 4 GB mobile hotspot data/line/mo.
$45 one-time activation fee (per plan)
**Data overage fee: If monthly 4 GB data limit is exceeded, a $10 fee for every 2 GB additional data applies.*

Number of Months	Total Cost
1	$95
4	$245
8	$445
12	$645

Phone Plan D

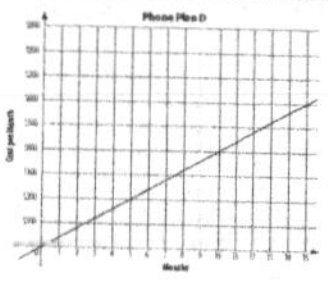

C = $40M

$40/Month! Nuwave Wireless
UNLIMITED talk, text, & data
* NO activation fee
* NO upgrade fee
* FREE unlimited mobile hotspot data!

Hotspot speed capped at 5 Mbps; limit 1 device tethered at a time

Number of Months	Total Cost
1	$40
3	$120
6	$240
12	$480

Activity 2: It Would Depend on the Person

Answers will vary. Sample answers:

1. Mary Jane should choose the Z-phone plan if money is less of a concern and she ends up using her hotspot to stream her videos on a different electronic device (like a laptop or smart TV) at some point. If she does not think she'll use a hotspot, she can save some money with the Cellular Wireless plan. The worst plan for her is Tel-Mobile because of the 4 GB of data and the additional $10 she will have to pay for another 2 GB of data if she goes over in a month.
2. Hassun should choose the Z-phone plan because of the 40 GB of hotspot data and unlimited talk and text. If internet service becomes available to him, his unlimited high-speed data will be an additional resource for him and his family to use without having to rely on his hotspot all the time. The worst plan for Hassun is the Cellular Wireless plan because it does not include any hotspot data. The Nuwave wireless plan is almost as bad, however, because it only allows one device to be tethered to a phone's hotspot at a time. Meaning, if Hassun uses his hotspot for his laptop, his family could not also videochat with family on a different device using the same hotspot.
3. Akari should choose the Nuwave Wireless plan because it is the cheapest plan, has no activation or upgrade fee, and has unlimited talk, text, and data. If she needs or wants to use her phone more often than just for emergencies, she will not have to pay anything more. The worst plan for Akari is the Z-phone plan because it is the most expensive and has more services than she needs if only used for emergencies.
4. Answers will vary.

Practice: I Am Changing the Rules!

Answers will vary. Sample answers:

The new equation: $C = \$70M$

The new table:

Number of Months	Total Cost
1	$70
3	$210
6	$420
12	$840

The new graph:

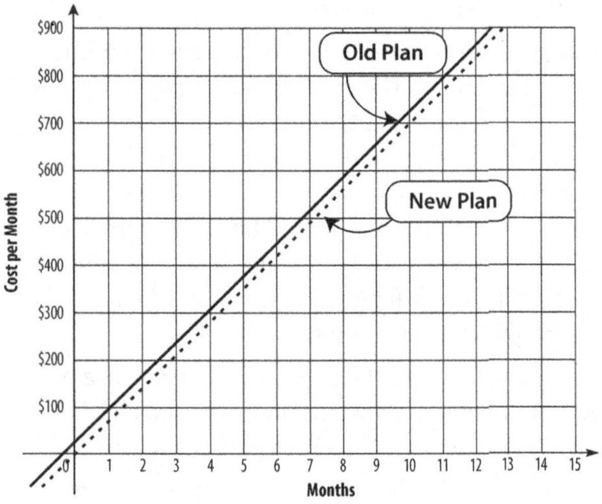

Extension: Looking at Four Graphs

Answers will vary. Sample answer:

Juania and others on her plan probably use their phones for more than just talking and texting. They probably use a lot of data each month and need it to be consistent in its speed without much buffering. They may be getting a better deal than what's on the graph with multiple lines using the same plan, so the lower costs for Plans C and D may be less of a concern, especially since those plans have more restrictions with data use and throttled speeds. Juana and others on her plan might look more closely at the differences between Plans A and B to determine which one gives them the lowest monthly payment but with the most features used by the group.

Blackline Master 7

Equations, Tables, and Graphs

Equation A

$y = 4x + 50$

Equation B

$y = 8x - 7$

Table C

In (x)	Out (y)
1	1
3	3
4	4
17	17

Table D

In (x)	Out (y)
3	65
7	85
10	100
11	105

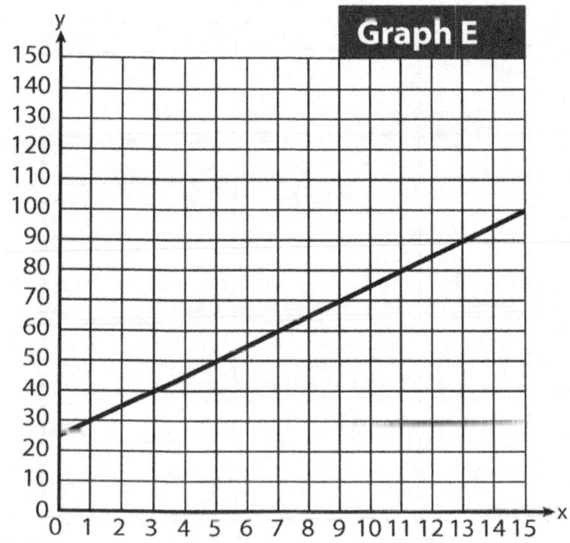

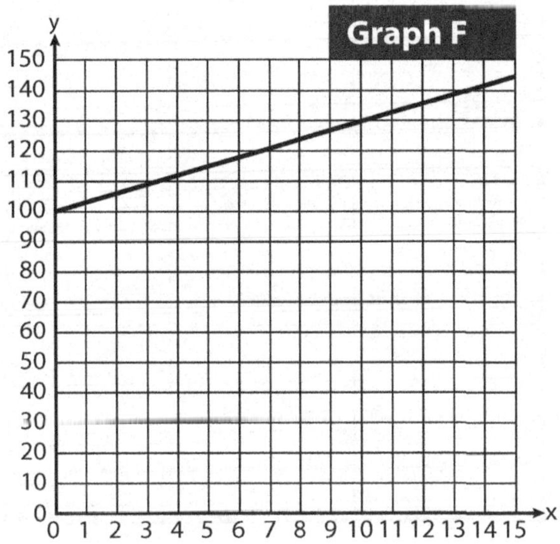

Blackline Master 8

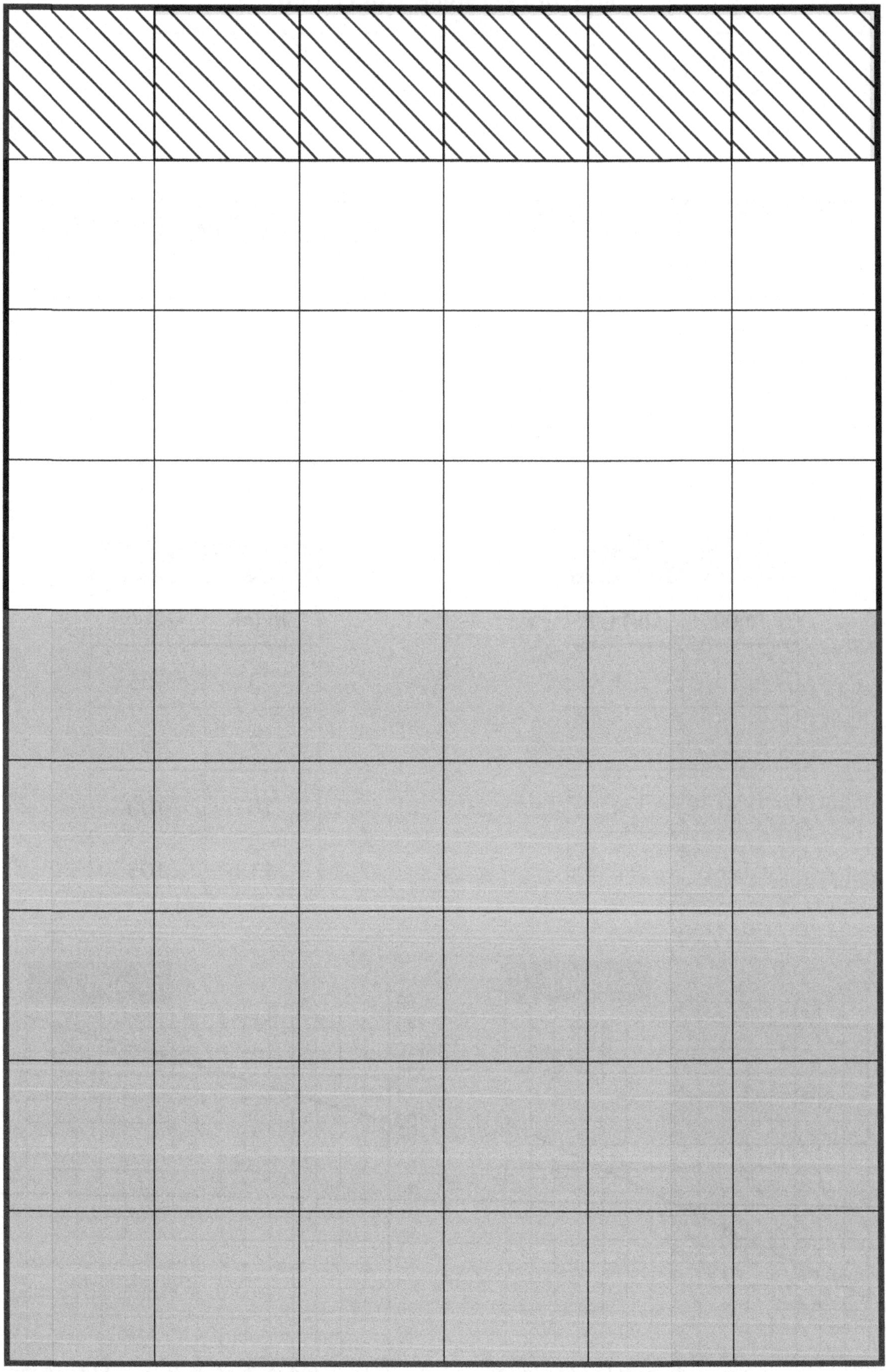

Blackline Master 9
Three Types of Tiles

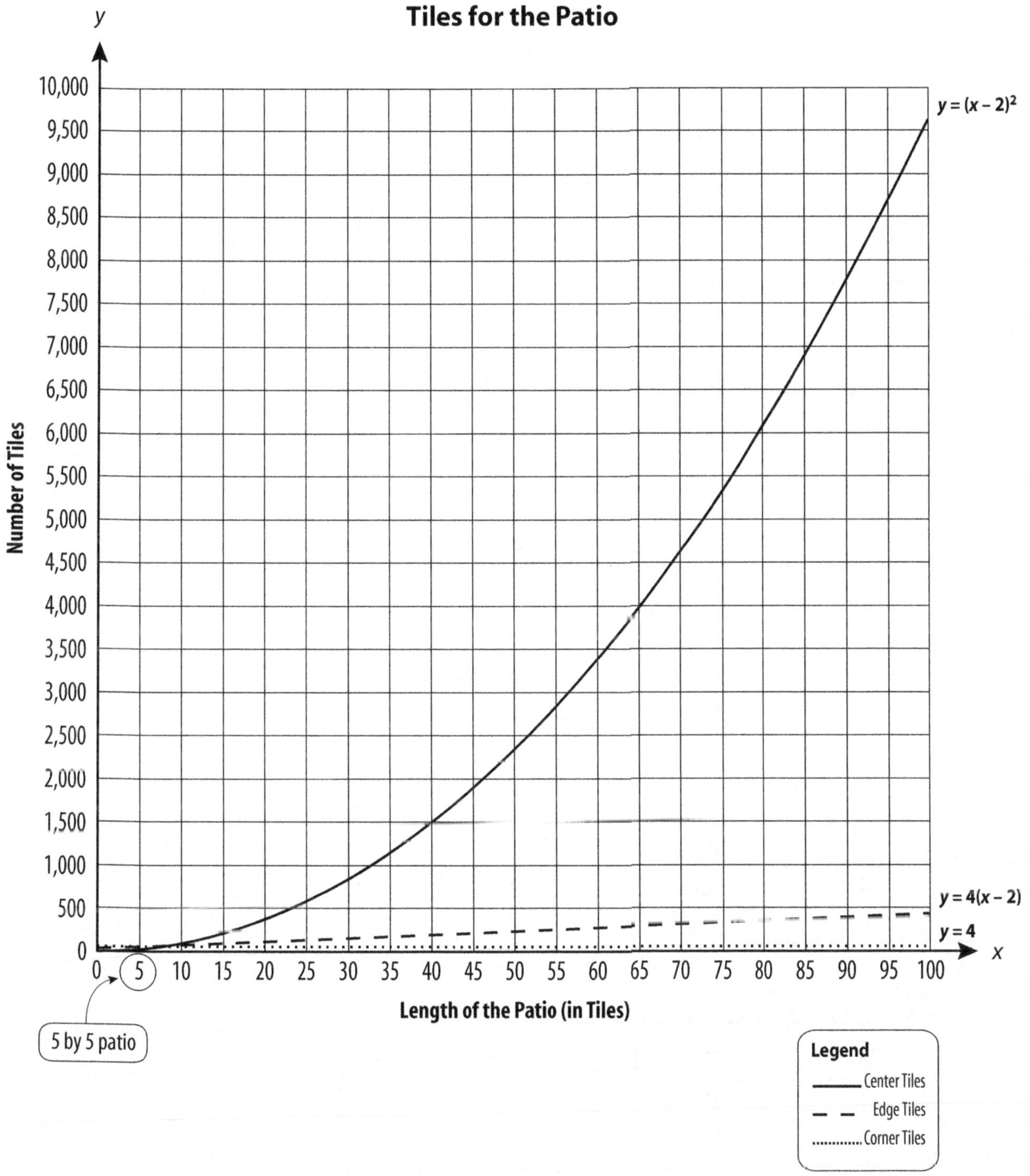

Blackline Master 10

Match Representations

x	y
1	6
2	12
3	18
4	24

x	y
1	40
2	40
3	40
4	40

$y = 6x$

x	y
1	10
2	40
3	90
4	160

x	y
1	86
2	92
3	98
4	104

$y = 10x^2$

x	y
1	20
2	40
3	60
4	80

x	y
1	35
2	50
3	65
4	80

$y = 10x$

x	y
1	10
2	20
3	30
4	40

$y = 40$

Blackline Master 11

Match Representations 2

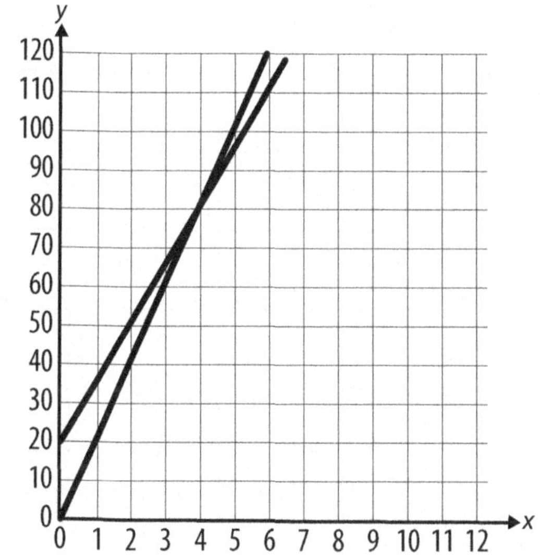

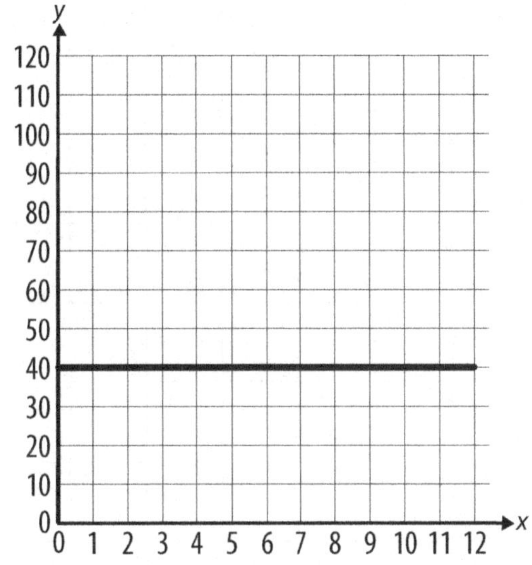

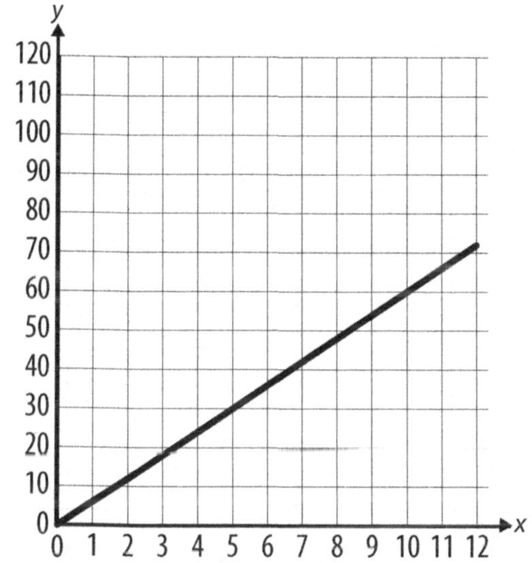

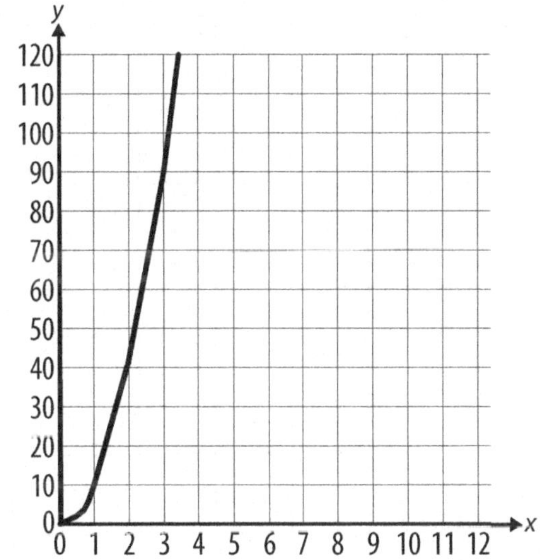

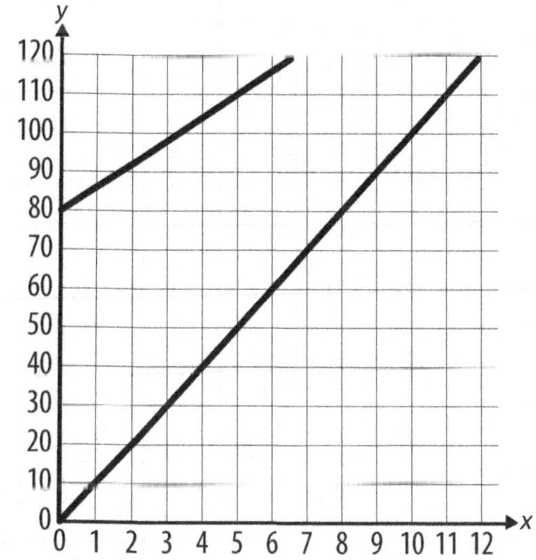

$y = 15x + 20$

$y = 20x$

$y = 6x + 80$

Blackline Master 12
1 cm Grid Paper

Blackline Master 13

0.75 cm Grid Paper

Blackline Master 14

Glossary

Reminder: Students generate their own definitions for terms as they arise in class, using language that makes sense to them. However, to help you guide the discussion, we include mathematical definitions for most of the terms.

Italicized terms are *not* included in the *Student Book*, but are included in the *Teacher Book* as background information. The lesson number in which the term first appears is in parentheses following the term.

circumference (6)—the distance around the outside of a circle.

coefficient (2)—the numerical factor of a term; for example, $2x$ means "2 times x" and 2 is the coefficient of x. Conventionally, the numerical coefficient precedes the variable.

constant (2)—a quantity whose value never changes.

constant function (10)—a function for which every ordered pair has the same second number; the graph of a constant function is a flat, horizontal line. The rate of change, or slope, has a value of zero.

constant rate of change (10)—a defining characteristic of linear patterns. Every time input (x) values increase by one, output (y) values increase by some constant amount. For example, in the equation $y = 3x$, y-values increase by three every time x-values increase by one. On a graph, the constant rate of change is made visible with steps in which you go over one and up three between successive points.

coordinate graph (4)—a grid formed by intersecting vertical and horizontal number lines, used for plotting points by ordered pairs.

coordinates (4)—the numbers in an ordered pair used to locate a point on a coordinate plane; the first number in the pair is usually called the x-coordinate and the second number, the y-coordinate.

dependent variable (4)—in a function with two variables, the dependent variable represents the output (y) value in each ordered pair. The values are determined by applying a rule to the input values.

diameter (6)—the distance through the center of a circle.

equation (1)—a mathematical sentence that indicates two quantities are equal; an equation always has an equals sign (=).

exponent (11)—a raised number to the right of another number that indicates how many times the base (unraised) number is to be used as a factor; for example, in the term 4^3, the three is the exponent or power used for shorthand in the equation $4 \times 4 \times 4 = 64$.

exponential pattern or **relationship** (11)—a function whose equation contains the variable as the power to which a constant is raised; for example, $y = 2^x$. The graph is curvilinear, and the rate of change varies along the graph.

flat-line graph (9)—represents a constant function where every point on the graph has the same y-value.

formula (6)—an equation showing a mathematical relationship in which the letters stand for specific quantities; for example, in the formula $A = lw$, A stands for area, l stands for length, and w stands for width.

function (1)—a relationship between two variables in which the value of one variable depends on the value of the other; if the variable y is a function of the variable x, then there is exactly one y-value for every x-value.

increments (4)—evenly spaced parts scaled to different amounts, also called intervals.

independent variable (4)—the variable that represents the input (x) values in each ordered pair in a function with two variables; output values are determined by applying a rule to each input value.

In-Out table (1)—a two-column table with a rule built into it that relates the Out values to the In values.

labeling points (4)—when points are plotted on a coordinate graph, they may be labeled with an ordered pair of numbers (x, y) to indicate their location on the plane. It is the convention to list the x-values first, the y-values second.

linear relationship (or **pattern**) (10)—a linear function in which the ordered pairs, when graphed on a coordinate plane, form a straight line.

line steepness (9)—indicates how fast the rate of change is occurring in a function; the greater the rate of change, the steeper the line graph.

multiple representations (Intro)—different forms for presenting information about a set of data, such as words, diagrams, tables, graphs, and equations.

nonlinear rate of change (11)—a rate of change that is *not* constant; nonlinear rates of change often appear as curves on a graph.

nonlinear pattern (or **relationship**) (11)—a nonlinear function in which the ordered pairs, when graphed on a coordinate plane, do not form a straight line.

ordered pair (4)—a pair of numbers in a specific order that can be used to plot a point on a coordinate graph, usually listed as (x, y).

origin (4)—the point where the x-axis and y-axis meet, labeled (0, 0).

pi (6)—represented by the symbol π; it represents the constant ratio of the circumference of a circle to the diameter, approximately 3.14.

point of intersection (8)—indicates a point where two graphs have the same coordinates; therefore, the two situations represented have identical values.

quadratic patterns or ***relationships*** (11)—a function whose equation contains at least one squared term; for example, $y = x^2 + 5$.

slope (10)—the steepness or slant of a line on a graph, measured by how much it rises or falls for each unit the line moves to the right.

table (1)—a chart with information arranged in rows and columns.

term (Intro)—a part of an algebraic expression that can be variable or constant; for example, $5x + 3y + 10$ has three terms; $5x$, $3y$, and 10. A term is also a member of a sequence of numbers arranged according to some pattern.

variable (1)—an amount that can have many values, which vary or change.

x-axis (4)—the horizontal line of a graph, usually marked in even intervals; the x-axis typically represents values for independent variables, such as time.

y-axis (4)—the vertical line of a graph, usually marked in even intervals; the y-axis typically represents values for dependent variables, such as wages earned over time.

y-intercept (8)—the point at which a graph touches or crosses the y-axis; for example, the point (0, 2) is the y-intercept for the function $y = x + 2$. The term is also used for the output value of a function when the input value is 0; for example, 2 is the y-intercept for the function $y = x + 2$.

Answer Key

Opening the Unit

Activity 1: Making a Mind Map

Answers will vary.

Activity 2: One of My Own Patterns

Answers will vary.

Activity 3: Initial Assessment: Patterns and Problems

Task 1: Visual Patterns

1. a.

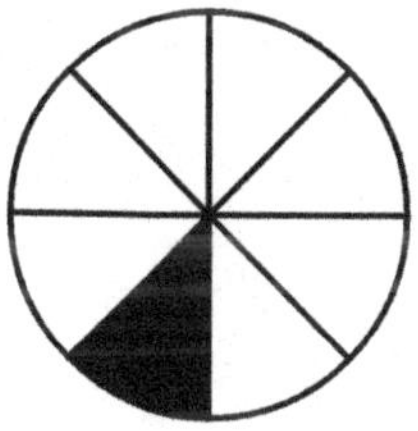

b.

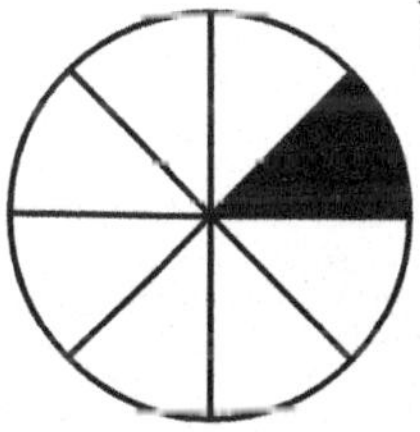

Answers will vary. Sample answer: I know what the 50th picture will look like because after 8, you begin the pattern again. 6 x 8 = 48, which means that the 49th would begin the cycle again.

2. e

Task 2: Number Patterns

x	y
2	11
3	14
5	20
10	35

Task 3: Make a Graph and a Rule

1. Answer will vary. Sample table:

Seconds	Hiccups
2	1
4	2
6	3
8	4

2.

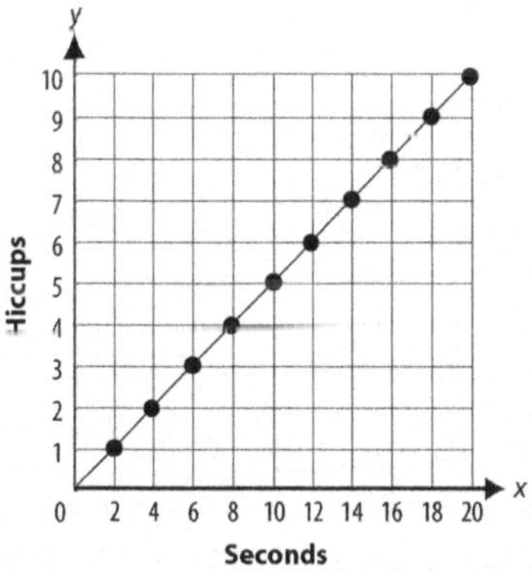

3. Rule: Every two seconds, another hiccup occurs, or divide the number of seconds by 2 to get the number of hiccups.
4. Equation: $y = x/2$

Task 4: Matching Graphs, Tables, and Equations

Graph A, Table 2, Equation 1

Graph B, Table 1, Equation 4

Graph C, Table 4, Equation 3

Graph D, Table 3, Equation 2

Task 5: Symbol Sense

1 **a.** False

b. False
Answers will vary. Sample answers:
$^{40}/_{10}$ π $^{10}/_{40}$, or $^{40}/_{10} > {^{10}/_{40}}$

c. False
Answers will vary. Sample answers:
$5 + 4(10)$ π 90, or $(5 + 4)(10) = 90$

d. True

e. True

2. a. 30

b. 54

c. 7

d. 6

Lesson 1: Guess My Rule

Activity 1: Guess My Rule

Answers will vary. Sample answers:

Table 1

x	y
2	6
3	9
4	12
5	15
6	18
7	21

The rule in words: Multiply x by 3 to get y.

The equation: $y = 3x$

Table 2

x	y
3	120
5	200
7	280
9	360
10	400
12	480

The rule in words: Multiply x by 40 to get y.

The equation: $y = 40x$

Table 3

x	y
56	7
96	12
104	14
80	10
144	18
8	1

The rule in words: Divide x by 8 to get y.

The equation: $y = {^x}/_8$

Table 4

x	y
1	7
18	24
21	27
34	40
12	18
60	66

The rule in words: Add 6 to x to get y.

The equation: $y = x + 6$

Table 5

x	y
2	0
3	1
4	2
14	12
20	18
200	198

The rule in words: Subtract 2 from x to get y.

The equation: $y = x - 2$

Table 6

x	y
1	3
2	5
4	9
8	17
10	21
12	25

The rule in words: Multiply x by 2, then add 1 to get y.

The equation: $y = 2x + 1$

Table 7

x	y
1	4
2	6
4	10
109	220
12	26
50	102

The rule in words: Multiply x by 2, then add 2 to get y, or Add 1 to x, then multiply by 2 to get y.

The equation: $y = 2x + 2$ or $y = 2(x + 1)$

Activity 2: Making Tables, Rules, and Equations

Answers will vary. See Activity 2, p. 20, for ideas on what to look for.

Practice: Guess More Rules

Entries in the table will vary. Sample answers:

Table 1

x	y
7	28
8	32
9	36
10	40

The rule in words: Multiply x by 4 to get y.

The equation: $y = 4x$

Table 2

x	y
3	24
5	40
6	48
8	64

The rule in words: Multiply x by 8 to get y.

The equation: $y = 8x$

Table 3

x	y
4	9
5	10
6	11
7	12

The rule in words: Add 5 to x to get y.

The equation: $y = x + 5$

Table 4

x	y
6	5
8	7
13	12
15	14

The rule in words: Subtract 1 from x to get y.

The equation: $y = x - 1$

Table 5

x	y
2	1.0
5	2.5
8	4.0
25	12.5

The rule in words: Divide x by 2 to get y.

The equation: $y = {}^{x}/_{2}$

Table 6

x	y
1	4
2	7
3	10
4	13
10	31

The rule in words: Multiply x by 3, then add 1 to get y.

The equation: $y = 3x + 1$

Table 7

x	y
3	22
4	29
11	78
20	141

The rule in words: Multiply x by 7, then add 1 to get y.

The equation: $y = 7x + 1$

Table 8

x	y
1	1
3	9
7	49
12	144

The rule in words: Multiply x by itself to get y.

The equation: $y = x^2$ or $y = (x)(x)$

Table 9

x	y
2	6
5	27
6	38
10	102
12	146

The rule in words: Multiply x by itself, then add 2 to get y.

The equation: $y = x^2 + 2$

Table 10

x	y
1	19
2	18
6	14
7	13
10	10

The rule in words: Subtract x from 20 to get y.

The equation: $y = 20 - x$

Practice: Fill in the Values

1. Rule: $y = 2x + 5$

x	y
1	7
2	9
3	11
4	13
5	15

2. Rule: $y = 25 - x$

x	y
12	13
11	14
16	9
4	21
5	20

3. Rule: $y = 3x - 1$

x	y
1	2
2	5
3	8
10	29
33	98

4. Rule: $y = x - 95$

x	y
100	5
101	6
102	7
200	105
295	200

5. Rule: $y = x^2 - 1$ or $y = (x)(x) - 1$

x	y
2	3
3	8
4	15
10	99
12	143

6. Rule: $y = x^2 - 5$ or $y = (x)(x) - 5$

x	y
5	20
10	95
12	139
15	220
20	395

Practice: The Rule Story

Terri managed an office where one of the supervisors was always 10 minutes late for work. This annoyed Terri, so she decided to make a **table** to keep track of the accumulated lost work time. "I'm going to show James how his **pattern** of lateness mounts up over **time**," she thought. The two things Terri recorded were the number of days James was late and the number of lost **minutes** that built up, or accumulated. On day one, Terri recorded that 10 minutes of work time had been lost. By the end of the workweek, her **table** showed that James had been late 10 minutes every workday and had lost a total of **50** minutes of work time.

Terri decided to confront James. "Look at this **table**," she said. "You can see that in a month (20 workdays), you will have cost this company **200** minutes of lost work time. That's more than **3** hours. And at your salary, that's a lot of money!"

James didn't believe Terri. "Show me how you figured that out," he said.

Terri agreed to explain how she could predict the time James would lose because of his **pattern** of lateness. "See here," she said. "You take the number of **minutes** late each day, always the same, and multiply it by the **number of days** you will be late to find out the total number of minutes you will have lost over the course of a month." Just to make her point, she wrote on a piece of paper the rule for James. That rule looked like this: **10 minutes multiplied by the number of days = minutes lost**.

James was paid $30 an hour. Even he could figure out that at that rate of pay, his lateness cost the company **100** dollars a month. He was ashamed of himself, but proud of his manger. "Terri," he said, "I always knew you were good at math. I just didn't know *how* good."

Symbol Sense Practice: How Many Ways?

Answers will vary. Sample answers:

Five times four

$5 \cdot 4$

5(4)

3 x 42

Three times forty-two

3(42)

39(84)

$39 \cdot 84$

Thirty-nine times eighty-four

ab

a times b

$a \cdot b$

lwh

l times w times h

$l \cdot w \cdot h$

$5 \cdot 10$

5 x 10

five times ten

$24x + 3$

Twenty-four times x plus 3

$24(x) + 3$

$b = 100 - 6a$

b = six times a subtracted from 100

$b - 100 - 6(a)$

$I = prt$

$I = p$ times r times t

$I = p(r)(t)$

Symbol Sense Practice: Four Ways to Write Division

Answers will vary. Sample answers:

$8\overline{)56}$	56 ÷ 8	$\frac{56}{8}$	56 divided by 8
$5\overline{)30}$	30 ÷ 5	$\frac{30}{5}$	**30 divided by 5**
$6\overline{)180}$	**180 ÷ 6**	$\frac{180}{6}$	**180 divided by 6**
$1\overline{)40}$	40 ÷ 1	$\frac{40}{1}$	**40 divided by 1**
$7\overline{)2}$	**2 ÷ 7**	$\frac{2}{7}$	2 divided by 7
$y\overline{)x}$	$x \div y$	$\frac{x}{y}$	***x* divided by *y***
$10\overline{)m}$	$m \div 10$	$\frac{m}{10}$	*m* divided by 10
$\$.50\overline{)\$10.00}$	**\$10.00 ÷ \$.50**	$\frac{\$10.00}{\$.50}$	**\$10.00 divided by \$.50**

Extension: Building Tables from Rules

Answers will vary. Sample answers:

$y = {}^1\!/_4 x$

x	y
4	1
8	2
20	5
100	25
200	50

$y = 2x + 0.75$

x	y
1	2.75
3	6.75
5	10.75
10	20.75
15	30.75

$y = {}^1\!/_2 x - 1$

x	y
2	0
4	1
10	4
20	9
25	11.5

$y = 1.8x$

x	y
1	1.8
2	3.6
5	9.0
10	18.0
25	45.0

Test Practice

1. (1)
2. (4)
3. (2)
4. (2)
5. (5)
6. 26

Lesson 2: Banquet Tables

Activity: Banquet Tables

Task 1

1. Arrangement 2: 6 Arrangement 3: 8
2. 10 people could be seated at it.

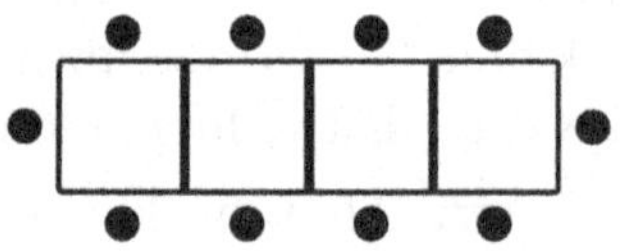

3. **b.** 22
   Explanations will vary. Sample answer: There would be 2 people seated at each of the 10 tables, plus there would be one person at each end of the group of tables. 10 x 2 + 2 = 22.
4. Explanations will vary. Sample answer: There would be 2 people at each of the 100 tables (200), plus one at either end of the tables (2) for a total of 202.

5. Sample entries:

Tables	People
1	4
2	6
3	8
4	10
10	22
100	202
200	402
250	502

6. The rule in words: There are always two people per table, plus one more at either end (for a total of 2 extra). Or put one person at each table and one at the end, then double that.

   The equation: $p = 2t + 2$, where p is the number of people and t is the number of tables.
   Or, $p = 2(t + 1)$.
7. Answers will vary. Sample answer:
   $2(200) + 2 = 202$.

Task 2

1. **c.** 9
   Explanations will vary. Sample answer: There would be 2 people at the ends of the group of tables, so that leaves 18 people sitting across from each other. Because there are 2 per table, 9 tables would be required.
2. Explanations will vary. Sample answer: First, you would subtract the 2 people that would sit at either end of the table ($100 - 2 = 98$). Then, you would divide 98 by 2, because there are two people per table (not counting the ends). 49.
3. Explanations will vary. Sample answer: Subtract 2 from the total number of people, then divide what is left by 2 ($t = (p - 2)/2$).
4. Answers will vary.
5. Explanations will vary. Sample answer: To find the total number of people, you add 2 after you multiply the number of tables by 2. To find the total number of tables, you subtract 2, then divide by 2.

Practice: Toothpick Row Houses

1. It takes 11 toothpicks for the second arrangement.
2. It takes 16 toothpicks for the third arrangement.
3. Arrangement 4

4.

Houses (n)	Toothpicks (t)
1	6
2	11
3	16
4	21

5. It takes 51 toothpicks for the 10 row houses. Explanations will vary. Sample answer: There are 5 toothpicks needed for every house except the first, which needs 6. Each of the 10 row houses needs 5 toothpicks except for the first, which needs one extra.
6. Explanations will vary. Sample answer: Five toothpicks are needed for each of the 100 row houses, except the first, which needs one extra (501).
7. **a.** Explanations will vary. Sample answer: For any number of houses, multiply that number by 5, then add 1. $t = 5n + 1$.
   b. Answers will vary. Sample answer:
   6 houses = $6(5) + 1 = 31$ toothpicks.
8. You could build 25 houses with 126 toothpicks. Explanations will vary. Sample answer: First, take away the one extra toothpick for the first house. That leaves 125 toothpicks. Because each row house takes 5 toothpicks, divide 125 by 5, which equals 25. $n = (126 - 1)/5$.

Symbol Sense Practice: Rules of Order

My Dear Aunt Sally

1. $5 - 4 + 6/2 = 4$
2. $120 \times 4 + 3 - 2 = 481$
3. $100(62) - 30(62) = 4{,}340$
4. $100/2 - 50/5 = 40$
5. $6 + 4(3) + 12 = 30$
6. $6 + 4(3)/12 = 7$

Parentheses

1. $10(1 + 20) = 210$
2. $(4 + 3) \times (2 + 5) = 49$
3. $(9 + 8 - 7)(6 + 5 - 3) - (2 + 1) = 77$
4. $10(6 - 1) + 5 = 55$
5. $10(6) - 1 + 5 = 64$

6. 10(6 – 1 + 5) = 100
7. 10(6) – (1 + 5) = 54

You Choose the Operations

Answers will vary. Sample answers:

1. (4 – 3) + (4 + 1) = 6
2. 4 + 3 – 4(1) = 3
3. (4 + 3) – (4 – 1) = 4

Extension: The Importance of Order

1. Answers will vary. Sample answer:

   Rule 1: $y = 5x + 2$

x	y
2	12
3	17
5	27
10	52
15	77

   Rule 2: $y = 5(x + 2)$

x	y
2	20
3	25
5	35
10	60
15	85

   Rule 3: $y = 5x + 10$

x	y
2	20
3	25
5	35
10	60
15	85

2. Explanations will vary. Sample answer: Rules 2 and 3 are the same because $5(x + 2)$ is the same as $5x + 10$. Rule 1 is not the same because you are only adding 2 to $5x$, not multiplying 5 by 2. Or adding 2 to any number and then multiplying the result by 5 gives the same result as adding 10 to 5 times the number. Rule 1 is not the same because you are only adding 2 to 5 times the number.

Test Practice

1. (2)
2. (3)
3. (3)
4. (2)
5. (2)
6. 15

Lesson 3: Body at Work—Tables and Rules

Activity 1: Heart Rates at Rest—Part 1

Name	Beats per 15 Seconds	Beats per Minutes
A baby	30	120
A feverish adult	35	140

Answers will vary. See *Activity 1*, p. 42, for student examples. Sample answers:

The rule in words: Multiply the number of beats per 15 seconds by 4 to get beats per minute.

The equation: $y = 4x$. y = number of beats per minute. x = number of beats per 15 seconds.

Heart Rates at Rest—Part 2

Answers will vary but should reflect accurate calculations. Sample answer based on heart rate of 80 beats per minute:

Time in Minutes (M)	My Total Number of Heartbeats (B)
1	80
5	400
10	800
60	4,800
1,440	115,200

1. The rule in words: Multiply the number of beats in one minute (80) by the number of minutes your heart beats.

   The equation: $B = 80M$

2. The rule in words: Divide the total number of beats by the number of beats your heart beats in one minute (80).

   The equation: $M = B/80$

3. Answers will vary. Sample answer based on 80 beats per minute: Between 8 and 9 days (921,600 beats in 8 days; 1,036,800 in 9 days).

Activity 2: How Many Calories Am I Burning?

Minutes Jogging	Calories Burned
1	10
15	150
30	300
45	450
60	600
100	1,000
120	1,200

Answers will vary. Sample answers:

Pattern: For every minute of jogging, 10 calories are burned.

The rule in words: To find calories burned while jogging, multiply the number of minutes jogged by 10.

The equation: C (Calories burned) = $10M$ (Minutes of jogging)

Minutes Cleaning House	Calories Burned
1	5
15	75
30	150
45	225
60	300
100	500
120	600

Answers will vary. Sample answers:

Pattern: For every minute of house cleaning, 5 calories are burned.

The rule in words: To find calories burned while cleaning house, multiply the number of minutes cleaning house by 5.

The equation: C (Calories burned) = $5M$ (Minutes of cleaning)

Minutes Running Up Stairs	Calories Burned
1	20
15	300
30	600
45	900
60	1,200
100	2,000
120	2,400

Answers will vary. Sample answers:

Pattern: For every minute of running up stairs, 20 calories are burned.

The rule in words: To find calories burned while running up stairs, multiply the number of minutes running up stairs by 20.

The equation: C (Calories burned) = $20M$ (Minutes of running up stairs)

Minutes Sitting	Calories Burned
1	1
15	15
30	30
45	45
60	60
100	100
120	120

Answers will vary. Sample answers:

Pattern: For every minute of sitting, 1 calorie is burned.

Rule in words: To find calories burned while sitting, multiply the number of minutes sitting by 1.

The equation: C (Calories burned) = $1M$ (Minutes of sitting), or $C = M$

1. **a.** 300 minutes

   Explanations will vary. Sample answer: You can look at the table for the number of calories burned while jogging for 30 minutes: 300. Then you can look at the pattern in the table for minutes sitting and see that you would have to watch TV for ten 30-minute periods to burn the same number of calories.

b. Answers will vary. Sample answer: The equation for calories burned while jogging is $C = 10M$; the equation for calories burned while sitting is $C = M$. Therefore, it takes 10 times longer to burn calories while sitting than jogging.

2. Answers will vary. Sample answers:

	Jogging	
	Min.	Cal.
Plan 1	180	1,800
Plan 2	170	1,700
Plan 3	120	1,200

	Cleaning House	
	Min.	Cal.
Plan 1	100	500
Plan 2	240	1,200
Plan 3	180	900

	Running Up Stairs	
	Min.	Cal.
Plan 1	60	1,200
Plan 2	30	600
Plan 3	70	1,400

	Total	
	Min.	Cal.
Plan 1	340	3,500
Plan 2	440	3,500
Plan 3	370	3,500

Rule to make a plan to burn 3,500 calories: You add the total number of calories for each type of activity so that they total 3,500. You multiply the number of minutes jogging by 10 to get the calories burned, the number of minutes cleaning house by 5 to get the calories burned, and the number of minutes running up stairs by 20 to get the calories burned.

Practice: Say It in Words and Fill in the Tables

Sample answers:

1. In words: For every minute, the cost is $0.15.

A table:

Dollars	Minutes
$0.30	2
$1.50	10
$4.50	30
$9.00	60

A possible situation: Cost of phone usage during off-peak times

2. In words: For every day, 2,500 calories are eaten

A table:

Calories	Days
5,000	2
12,500	5
17,500	7
35,000	14

A possible situation: Dietary intake of calories

Practice: Driving at 50 Miles per Hour

1. Answers will vary. Sample answer: It means that for every hour, she would travel 50 miles.

2. Answers will vary. Sample answer: No, but the rate stays the same; so if she drives only 30 minutes, she would travel 25 miles.

3. a. 1 hour? **50 miles**

b. 2 hours? **100 miles**

c. Half an hour? **25 miles**

d. 10 hours? **500 miles**

4. a. 50 miles? **1 hour**

b. 100 miles? **2 hours**

c. 25 miles? **0.5 hour**

d. 500 miles? **10 hours**

5. Explanations will vary. Sample answer: For problem 3, I multiplied the number of hours by 50. For problem 4, I divided the number of miles by 50.

6. **Driving at 50 mph**

Time (t)	Distance (d)
1	50
2	100
0.5	25
10	500

7. $d = 50t$ and $t = {}^{d}/_{50}$

Symbol Sense Practice: Equations ⟷ Words

Sample answers:

1. Add 9 to x to get y.
2. Multiply x by 10, then add 20 to get y.
3. Divide x by 8, then add 15 to get y.
4. To get y, take one-half of x and add 1 to it.
5. $y = 5x$
6. $2x = y$
7. $10x + 5 = y$
8. $y = x - 4$
9. Answers will vary.
10. Answers will vary.

Symbol Sense Practice: Substituting for *x*

Original Equations	Substitution Values for x $x = 0$	$x = 10$	$x = 100$
1. $y = 2x + 35$	35	55	235
2. $y = 15 + 3x$	15	45	315
3. $y = 5 + 7x$	5	75	705
4. $y = 2(x + 2)$ 4	24	204	
5. $y = (0.25 + 0.75)x$	0	10	100
6. $y = 1{,}000 + 25x$	1,000	1,250	3,500
7. $y = {}^{x}/_{2} + 90$	90	95	140

Extension: A Friendly Reunion

Table entries will vary. Sample answers:

1. a. **Sports Car Driver**

Time	Distance
1 hour	90
2 hours	180
3 hours	270
4 hours	360
5 hours	450
6 hours	540

Minivan Driver

Time	Distance
1 hour	60
2 hours	120
3 hours	180
4 hours	240
5 hours	300
6 hours	360

Old Pick-Up Truck Driver

Time	Distance
1 hour	45
2 hours	90
3 hours	135
4 hours	180
5 hours	225
6 hours	270

b. The sports car driver will need to leave by 10:00 a.m. The minivan driver will have to leave by 9:00 a.m. The pick-up truck driver will have to leave by 8:00 a.m.

2. Sample answer: The sports car driver was 2 hours late because she had to drive a total of 360 miles (90 + 90 + 180). The minivan driver was 1 hour late because she had to drive an extra 60 miles. The pick-up truck driver arrived on time at noon.

Test Practice

1. (5)
2. (2)
3. (2)
4. (4)
5. (2)
6. 6.25

Lesson 4: Body at Work—Graphing the Information

Activity: Four Graphs for Calories Burned

Graphs will vary depending on size of the intervals. Sample answers:

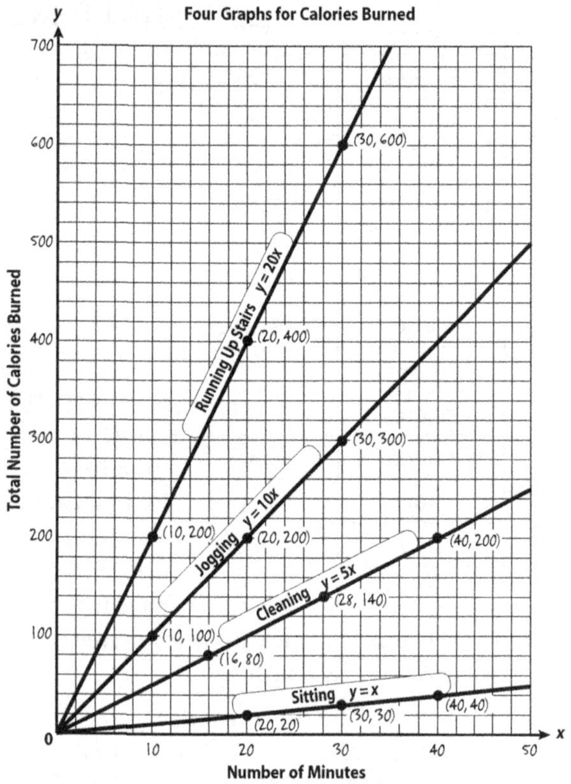

1. Similarities: They all increase over time. They are all straight lines.
2. Differences: The slant or slope of each of them is different.

A Closer Look at Four Graphs for Calories Burned

1. Answers will vary. Sample answer: The four lines intersect at the origin (0,0). They intersect here because that is when they are all the same. No minutes, no calories burned.
2. The "Running Up Stairs" graph has the steepest line because that exercise burns calories the fastest. The calories burned accumulate at a faster rate.
3. The "Sitting" graph has the flattest line because that exercise burns calories the slowest.
4. **a.** 2 hours. Indicated by the point (60,600) on the "Jogging" line and the point (120,600) on the "Cleaning" line.

   b. 855 calories. Indicated by the distance between (45,45) and (45,900).

   c. Yes, because 1,400 = 70(20).
5. Explanations will vary. Sample answer: The graph will start at (0,0) and slope upward as a straight line. The slope will be steeper than the "Cleaning" graph but less steep than the "Jogging" graph.
6. Sample table:

Minutes Shoveling Snow	Calories Burned
1	8
15	120
30	240
45	360
60	480
100	800
120	960

7. Explanations will vary. Sample answer:

   The rule in words: To find calories burned while shoveling snow, multiply the number of minutes shoveling by 8.

   The equation: C (Calories burned) = $8M$ (Minutes of shoveling)

Practice: Setting Up a Coordinate Graph

1. **a.** Coordinate graphs have two **axes**, which meet at the **origin**. The vertical axis is called the **y-axis**. The horizontal axis is called the **x-axis**. The scales on the axes should have **equal** increments.

   b.

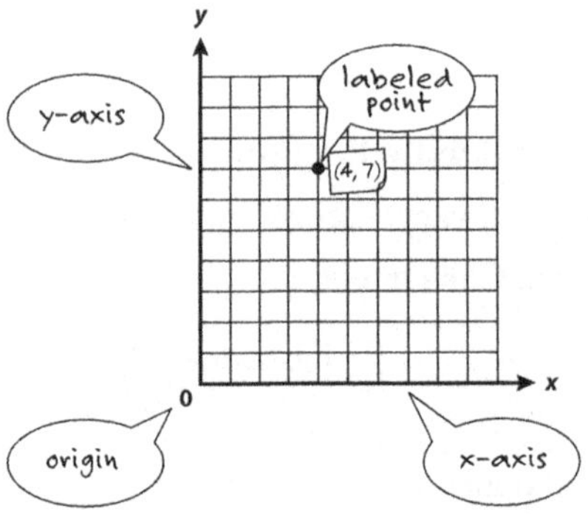

Practice: Graphing Guess My Rule

1.

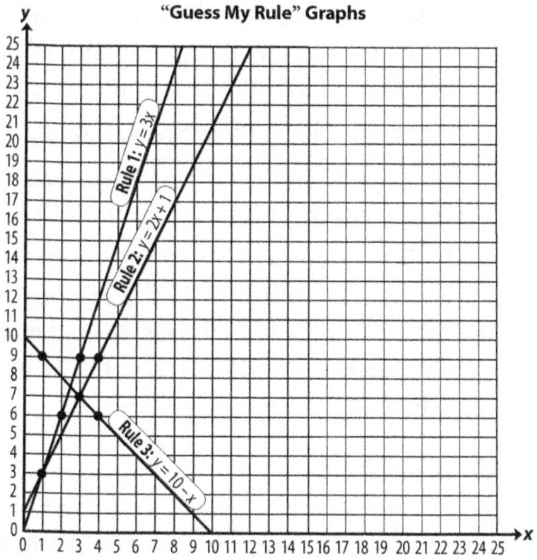

2. Answers will vary. Sample answers:

   Rule 1: Graph description: It is a straight line that begins at 0. It has a steep upward slope; *y* increases as *x* increases.

   Rule 2: Graph description: It is a straight line that begins at (0,1). It has a steep upward slope, although not quite as steep as rule 1; *y* increases as *x* increases.

   Rule 3: Graph description: It is a straight line that begins at 10. It has a downward slope, less steep than rule 2; *y* decreases as *x* increases.

Practice: Graphing Banquet Tables

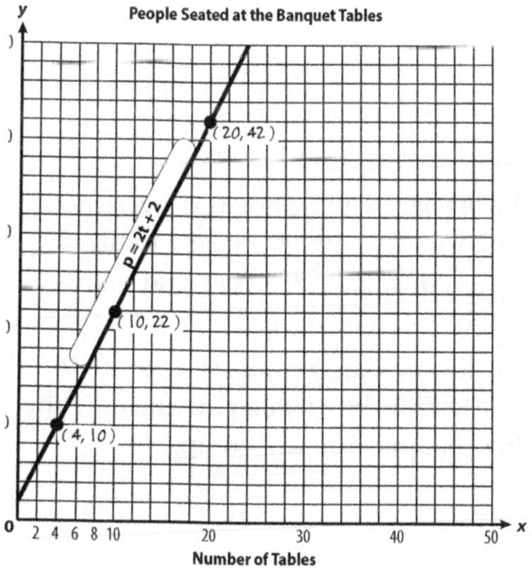

Symbol Sense Practice: Relating Multiplication and Division

1. 15 x 20 = 300

   $^{300}/_{15} = 20$

   $^{300}/_{20} = 15$

2. 1,000 = 40(25)

   $^{1,000}/_{40} = 25$

   $^{1,000}/_{25} = 40$

3. $3x = y$

   $^{y}/_{x} = 3$

   $^{y}/_{3} = x$

4. $5m = c$

   $^{c}/_{m} = 5$

   $^{c}/_{5} = m$

5. $c = 20m$

   $^{c}/_{m} = 20$

   $^{c}/_{20} = m$

6. $m = 60h$

   $^{m}/_{h} = 60$

   $^{m}/_{60} = h$

7. 6(40) = 240

Test Practice

1. (3)
2. (3)
3. (1)
4. (2)
5. (3)
6.

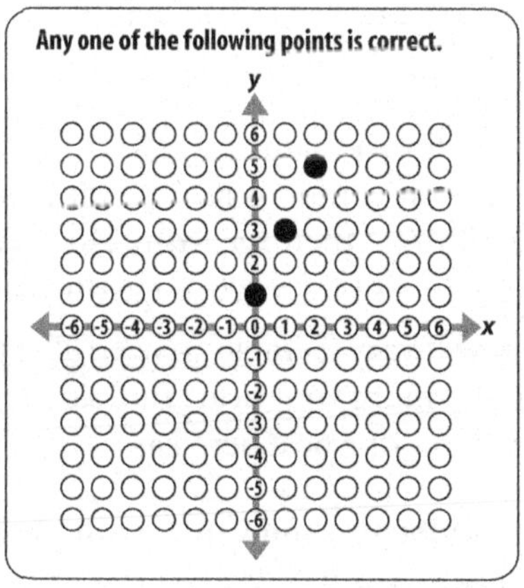

Lesson 5: Body at Work—Pushing It to the Max

Activity 1: Average Maximum Heart Rates

1. The average maximum heart rate for a person at age 75 should be 145.
2. The average maximum heart rate for a child at birth should be 220.
3. Answers will vary.
4. Answers will vary. See *Teacher Book, Activity 1*, p. 66, for student examples. Sample answer: Down the column, the pattern is that the age increases by 5 years, and the average maximum heart rate decreases by 5 beats.
5. Answers will vary. See *Teacher Book, Activity 1*, p. 66, for student examples. Sample answer: Across the rows, the pattern is that the age (number of years) and the AMHR (number of beats) always total 220.
6. 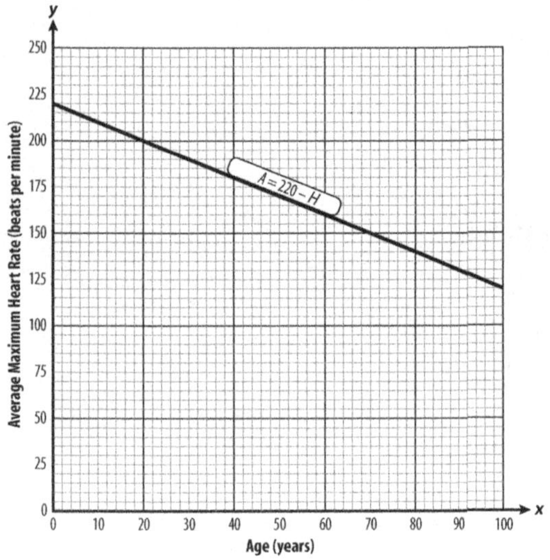

7. Answers will vary. Sample answer: To predict the average maximum heart rate for a person of any age, subtract their age from 220. Average maximum heart rate = 220 – age (in years).
8. Answers will vary. Sample answer: The older you get, the more your heart rate slows, on average about one beat per year.

Activity 2: Comparing and Contrasting Situations

Answers will vary. See *Teacher Book, Activity 2*, p. 69, for student examples. Sample answers:

1. **a.** Similarities: Both are straight lines.
   b. Differences: The Calorie line slants up; the AMHR line slants down.
2. **a.** Similarities: Both have a pattern.
   b. Differences: In the Calorie table, both numbers increase; in the AMHR table, as the age increases the heart rate decreases.
3. **a.** Similarities: Both have two variables.
   b. Differences: In the Calorie equation, you have to multiply, but in the AMHR equation, you have to subtract from 220.
4. Sample answer: Over time in the calorie pattern, the number of calories burned goes up. Over time in the AMHR pattern, the heart rate does down.

Practice: Brick Piles

Time of Day	Backyard Brick Pile (*B*)	Front-Walk Brick Pile (*F*)
8:00 a.m.	950	50
9:00 a.m.	900	100
10:00 a.m.	850	150
11:00 a.m.	800	200
12:00 p.m.	750	250
1:00 p.m.	700	300
2:00 p.m.	650	350
3:00 p.m.	600	400
4:00 p.m.	550	450
4:30 p.m.	525	475
5:00 p.m.	500	500

1. 525 bricks in the backyard
   475 bricks in the front walk
2. Answers will vary. Sample answer: Every hour 50 bricks moved from the backyard to the frontwalk. So, in a half hour, half that many bricks, or 25, would be moved.
3. Answers will vary. Sample answer: I think the backyard brick pile line will go down while the frontwalk brick pile line will go up.

4.

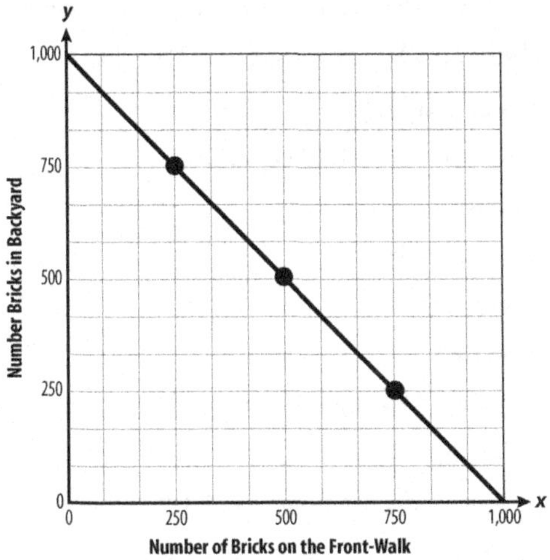

5. $F = 1{,}000 - B$ or $F + B = 1{,}000$

Symbol Sense Practice: Matching Equations

1. E
2. G
3. D
4. F
5. B
6. C
7. A
8. H
9. a. Answers will vary. Sample answer: If b = 5, then a = 0.
   b. Answers will vary. Sample answer: If b = 12, then a = 14.
10. Answers will vary. Sample answer.
    Subtraction Equation: $204 - x = y$
    Addition Equation: $204 = y + x$

Extension: Aerobic Target Heart Rate

Answers will vary.

Test Practice

1. (3)
2. (3)
3. (2)
4. (3)
5. (5)

6.

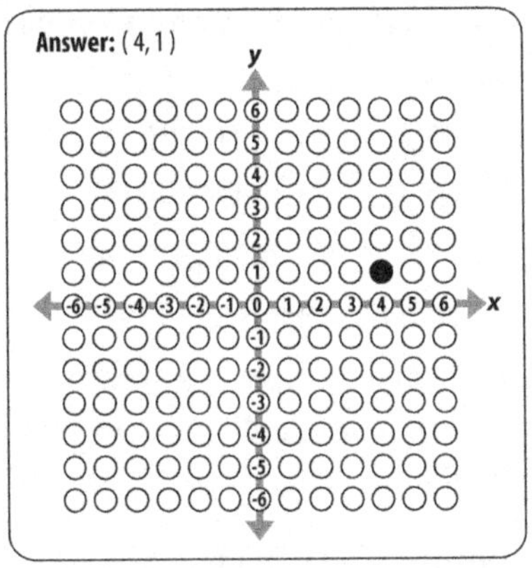

Lesson 6: Circle Patterns

Activity 1: Journey Through the Center of the Earth

1–6. Answers will depend on data collected by the students. Look at *Activity 1*, p. 79, for actual student examples.

The main point is that students observe and conclude that no matter the size of the circle, the circumference is always a little more than 3 times longer than the diameter and that $C \approx 3d$.

7. Answers will vary. Sample answer: The Earth's diameter is between 7,500 and 7,800 miles, I think, because 23,700 ÷ 3 = 7,900. The diameter seems to be a little less than this, because the circumference is a little more than 3 times longer.
8. Answers will vary. Sample answer:

Diameter	Circumference
2 inches	6 inches
3 inches	9.5 inches
5 inches	15.5 inches
6 inches	19 inches
About 7,500–7,900 miles	23,700 miles

Activity 2: Measure, Calculate, and Compare

Sample table entries:

1.	2	6	6.28	6.283185307
2.	3	9.5	9.42	9.424777961
3.	5	15.5	15.7	15.70796327
4.	6	19	18.84	18.84955592

5. Answers will vary. Sample answer: The more exact the estimate of *pi*, the more exact the circumference.

Practice: Graphing Circle Data

1. Answers will depend upon the data collected. Sample answer:

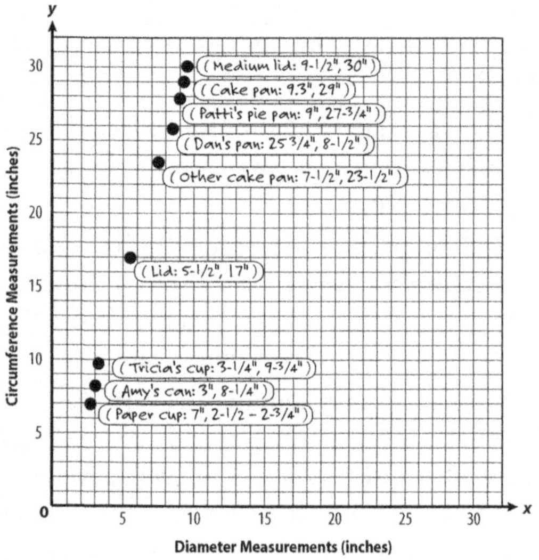

2. Answers will vary. Sample answer: As the diameter increases, so does the circumference.

3. The diameter would measure about 2.2 inches. Plotted point is (2.2,7).

4. The circumference would measure about 22 inches. Plotted point is (7,22).

Practice: How Much More Rubber?

About 9 inches. Sample explanation: Because the circumference of a circle is just a little more than three times the diameter, then a 24-inch tire would have a circumference of a little more than 72 inches (75.36). The circumference (or length) of the 27-inch tire is more 81 inches (84.780). The difference between 81 inches and 72 inches is 9 inches (84.78 – 75.36 = 9.42).

Symbol Sense Practice: = or ≈ or < or > or π?

1.

When You Want to Say …	Use This Symbol
… is exactly equal to …	=
… is approximately equal to …	≈
… is not equal to …	π
… is more than …	>
… is less than …	<

2. **a.** True

b. False $54.00 = $44.00 + $10.00

c. True

d. False $3 \approx 3.14$ (or $3 < 3.14$)

e. True

f. False $4x = x + x + x + x$

g. False $\frac{1}{3} > \frac{1}{4}$ (or $\frac{1}{3}$ π $\frac{1}{4}$)

h. False $\pi \approx 3$ (or $\pi > 3$)

Extension: Investigating Another Pattern in Geometry

1.

Polygon Name	Number of Sides	Sum of the Angle Measures (degrees)
triangle	3	180
square	4	360
pentagon	5	540
hexagon	6	720
octagon	8	1,080

2. **b.** $S = 180(n - 2)$

3. Answers will vary. Sample answer: There are 6 sides to a hexagon. I used the rule $S = 180(6 - 2)$, which is the same as 180(4) or 720 degrees.

Test Practice

1. (4)

2. (2)

3. (4)

4. (2)

5. (4)

6. 31.4

Interim Assessment: Using the Tools of Algebra

Task 1: Working from a Graph

1. Answers will vary. Sample answer: The cost per night at the hotel is $100. It costs the same, whether you stay one night or several.
2.

Nights Lodging	Cost
1	$100
2	$200
3	$300
4	$400
5	$500
6	$600

3. The next set of numbers on the table could be 7 nights lodging for $700.
4. The total cost for the 10th night would be $1,000. The total cost for the 20th night would be $2,000. The total cost for the 100th night would be $10,000. To figure either of these total accumulated costs, multiply the number of nights by $100.
5. a. The rule in words: Total accumulated cost equals $100 multiplied by the number of nights lodging.
   b. The equation: $C = 100n$, where C is the total cost and n is the number of nights lodging

Task 2: Working from a Table

1. Answers will vary. Sample answer: During a drought, the depth of the reservoir fell. For each week of drought, the water level dropped one foot.
2.

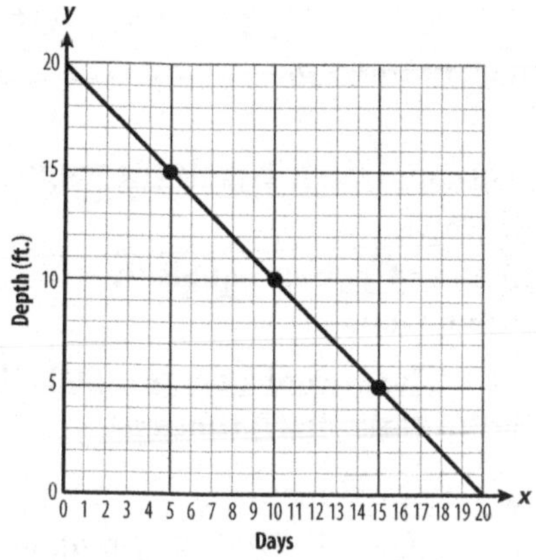

3. The reservoir depth on the 10th week will be 10 feet and 5 feet on the 15th week.
4. By the 20th week, the reservoir will run out of water. The depth started at 20 feet, and each week it loses a foot of depth.
5. a. The rule in words: To figure out the depth of the reservoir for any week, you subtract the number of drought weeks from 20.
   b. The equation: $d = 20 - w$, where d is the depth and w is the number of weeks of drought

Task 3: Working from an Equation

1. c
2. Table entries will vary.

Age (months)	Vocabulary Size
20	300
25	600
30	900
40	1,500
50	2,100

3.

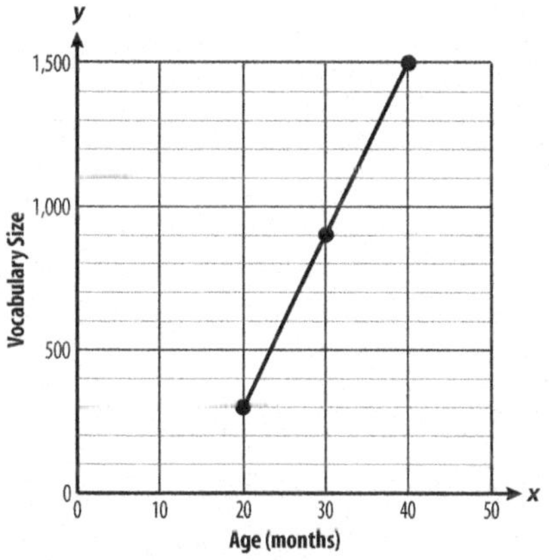

Task 4: Tina's Walking Rate

1. c

Task 5: Symbol Sense

Answers will vary. Sample answers:

2. ~~20m = 2 + m~~ 20m = 20 x m
3. ~~22 – 4(5) = 90~~ 22 – 4(5) = 2
4. ~~10(6 + 1) = 17~~ 10(6 + 1) = 70
5. ~~6 + 10 > 7 + 10~~ 6 + 10 < 7 + 10
6. ~~If x = 7, then 50 + 100x = 1050~~
   If $x = 7$, then $50 + 100x = 750$.
7. 900 = 700 + 200
8. $a = 30/b$ (or $b = 30/a$)
9. Multiply a number by 4, then add 2 for a total

of 16

10. $y = x + 6$

Lesson 7: What Is the Message?

Activity 1: Pass the Message

Answers will vary.

Activity 2: What Is the Message?

Answers will vary.

Sample answers for a multiplication equation such as $y = 12x$:

1. Table of $y = 12x$

x	y
1	12
2	24
3	36
4	48

2.

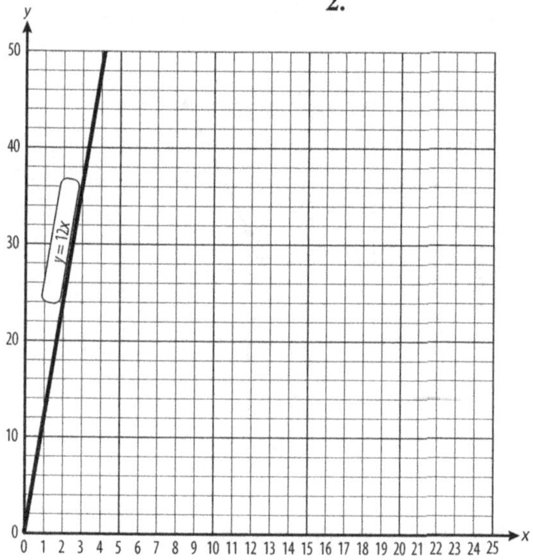

3. x is the number of cartons of eggs. Because there are twelve eggs in a carton, y represents the total number of eggs I have.

Sample answers for an addition equation such as $y = x + 12$:

1. Table of $y = x + 12$

x	y
1	13
2	14
3	15
4	16

2.

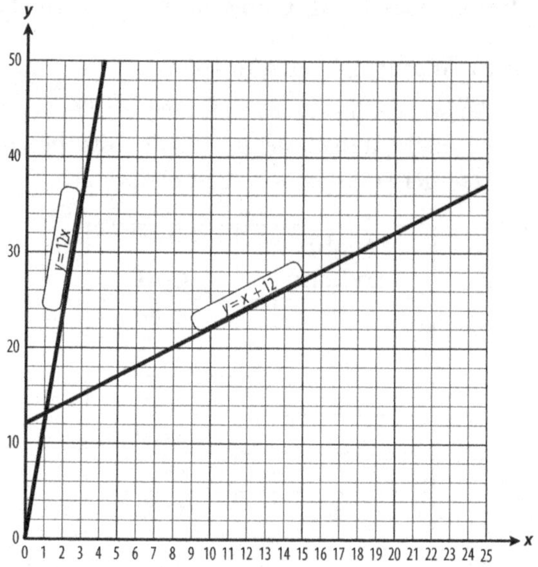

3. x is the amount of money I started with in my "pin money" account. y is the money that I have in my account now.

Practice: Geometry Formulas

Answers will vary. Sample answers:

$V = lwh$

The volume of a rectangular solid equals the length times the width times the height.

$P = 4s$

The perimeter of a square equals four times the length of a side.

$A = s^2 = s \times s$

The area of a square equals the length of a side times itself.

$V = s^3 = s \times s \times s$

The volume of a cube equals the length times the width times the height.

$C = 2\pi r$

The circumference of a circle equals the length of the radius times two times *pi*.

$C = \pi d$

The circumference of a circle equals the length of the diameter times *pi*

$d = C/\pi$ **The diameter of a circle equals the circumference divided by *pi*.**

$A = \pi r^2 = \pi \times r \times r$ **The area of a circle equals the length of the radius times itself times *pi*.**

Answers will vary. Sample answers:

Perimeter: **The distance around a rectangular object.**

Area: **The space inside a flat shape like a circle or rectangle.**

Volume: **The space inside a three-dimensional shape like a cube.**

Circumference: **The distance around a circle.**

Diameter: **The distance from one side of a circle to the other going through the center.**

Radius: **Half the distance of the diameter.**

Symbol Sense Practice: Evaluating Geometric Formulas

1. $A = 40(60) = 2{,}400$
2. $P = 2(90) + 2(10) = 180 + 20 = 200$
3. $A = 10 \times 10 = 100$
4. $P = 4(10) = 40$
5. $V = 20 \times 18 \times 5 = 1{,}800$
6. $V = 10 \times 10 \times 10 = 1{,}000$
7. $P = 2(100 + 62) = 2(162) = 324$
8. $C = 3.14 \times 10 = 31.4$
9. $A = 3.14 \times 10 \times 10 = 314$
10. $C = 2 \times 3.14 \times 10 = 62.8$
11. $A = 3.5 \times 6 = 21$
12. $P = 2(12.5) + 2(7.5) = 25 + 15 = 40$

Test Practice

1. (2)
2. (1)
3. (3)
4. (1)
5. (2)
6. 50

Lesson 8: Job Offers

Activity: Job Offers

Would They Ever Be the Same?

Answers will vary. See *Lesson in Action* for actual student examples of tables and graphs. The conclusion is that at the 40th week, the accumulated wages will be the same for both jobs ($8,000).

1. Sample table:

LaserLink	Weeks	QuinStar
$ 200	1	$2,150
$ 400	2	$2,300
$ 800	4	$2,600
$ 1,600	8	$3,200
$ 4,800	24	$5,600
$ 6,000	30	$6,500
$ 8,000	40	$8,000
$10,400	52	$9,800

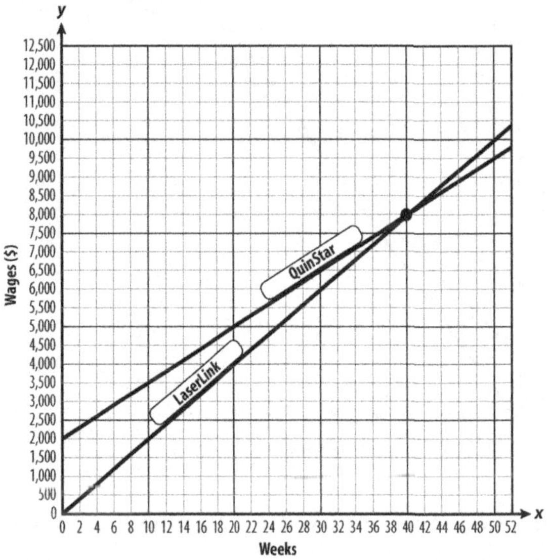

2. LaserLink rule in words: Multiply x (number of weeks) by 200 to get y (total pay).

   LaserLink equation: $y = 200x$
3. QuinStar rule in words: Multiply x (number of weeks) by 150, then add 2,000 to get y (total pay).

   QuinStar equation: $y = 150x + 2{,}000$
4. Answers will vary. Sample answers:
   a. What stays the same is the amount that gets multiplied weekly (200 for LaserLink and 150 for QuinStar).
   b. What changes is the amount of total pay each gets over time.
   c. What stands out for me is that QuinStar looked like a better deal at the beginning, but over time LaserLink wound up being a better job offer.

Practice: The Race

Answers will vary. Sample answers:

Jackie will catch up with Calvin at mile 8 and will win the race.

Miles Clavin Runs	Hours	Miles Jackie Runs
2	0	0
5	1	4
8	2	8
11	3	12

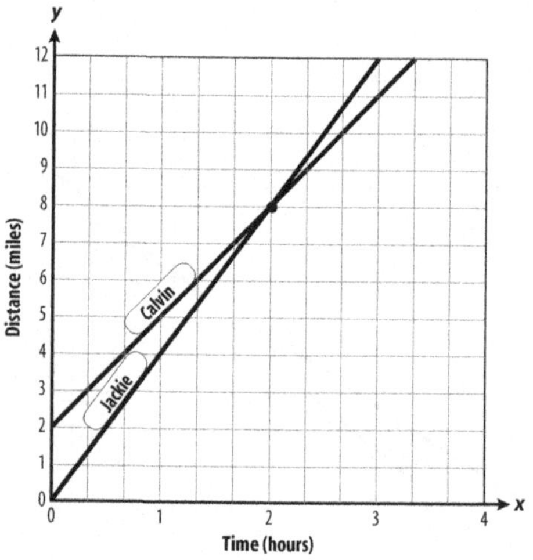

Practice: Armand's Weeks, Not Pay

1. 14,000 = 200w, so w = 70 weeks
2. 10,800 = 200w, so w = 52 weeks
3. $w = {}^{p}/_{200}$
4. 17,000 = 150w + 2000, so w = 100 weeks
5. 10,400 = 150w + 2000, so w = 56 weeks
6. $w = (p - 2{,}000)/150$

Symbol Sense Practice: Solving One-Step Equations

1. $x = 17$
2. $x = 1{,}010$
3. $x = 200$
4. $x = 150$
5. $x = 25$
6. $x = 100$
7. $x = 6$
8. $x = 4\frac{1}{2}$
9. $x = 21$
10. $x = 880$
11. $x + 7 = 20; x = 13$
12. $x + x = 100; x = 50$
13. $x + 100 = 6{,}875; x = 6{,}775$
14. $x - x = 0$; x = any number
15. $8x = 56; x = 7$
16. $x(x) = 100$ or $x^2 = 100; x = 10$
17. $2x = 1{,}200; x = 600$
18. $27 - x = 10; x = 17$
19. ${}^{x}/_{2} = 75; x = 150$
20. ${}^{x}/_{10} = 100; x = 1{,}000$

Extension: What If?

Answers will vary. Sample answers:

1. Until week 60, QuinStar would be the better job offer. At week 60, both jobs would have paid $12,000.

LaserLink	Weeks	QuinStar
$ 200	1	$ 3,150
$ 400	2	$ 3,300
$ 800	4	$ 3,600
$ 1,600	8	$ 4,200
$ 4,800	24	$ 6,600
$ 6,000	30	$ 7,500
$ 8,000	40	$ 9,000
$10,400	52	$10,800
$12,000	60	$12,000

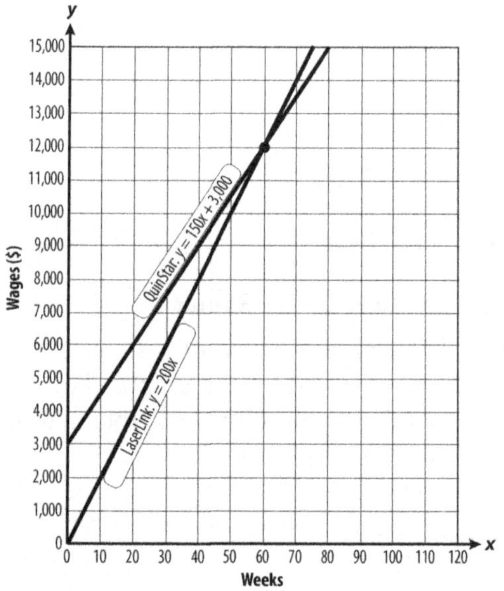

2. Without a bonus offer, LaserLink would be the better job offer. From the beginning, LaserLink pays $50 more per week.

LaserLink	Weeks	QuinStar
$ 200	1	$ 150
$ 400	2	$ 300
$ 800	4	$ 600
$ 1,600	8	$1,200
$ 4,800	24	$3,600
$ 6,000	30	$4,500
$ 8,000	40	$6,000
$10,400	52	$7,800

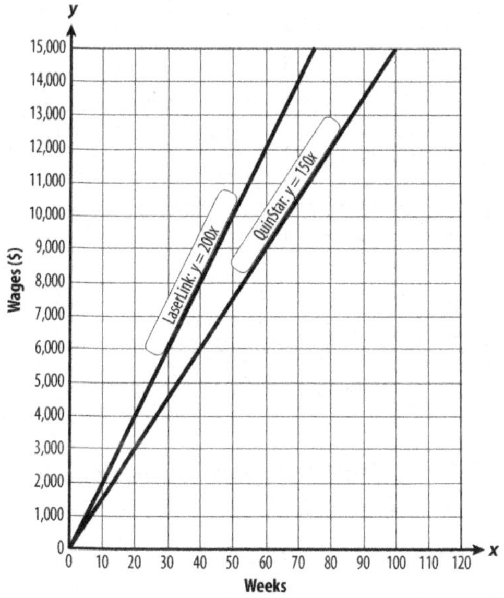

3. Clearly, LaserLink would be the better job offer.

LaserLink	Weeks	QuinStar
$ 2,200	1	$ 150
$ 2,400	2	$ 300
$ 2,800	4	$ 600
$ 3,600	8	$ 1,200
$ 6,800	24	$ 3,600
$ 8,000	30	$ 4,500
$10,000	40	$ 6,000
$12,400	52	$ 7,800

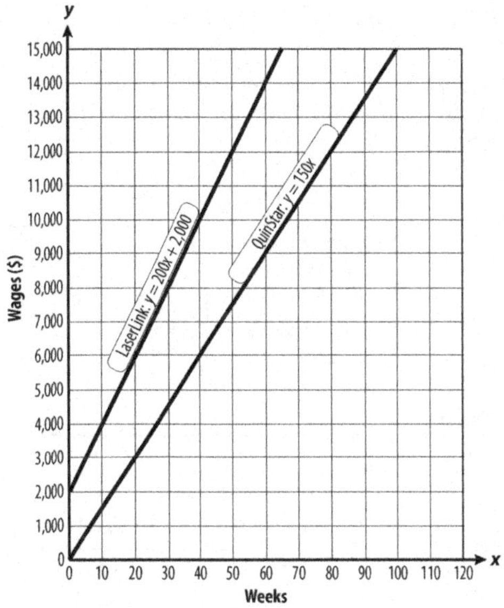

Test Practice

1. (4)
2. (3)
3. (3)
4. (4)
5. (5)
6. 150

Lesson 9: Phone Plans

Note: Answers for the updated version of this lesson can be found on the last page of Blackline Master 6a.

Activity 1: Phone Plans

Phone Plan A

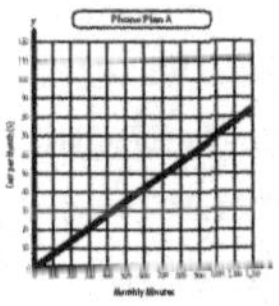

C = $0.07M

Cellular

7 cents per minute

No Monthly Charge!

Monthly Minutes	Cost
50	$3.50
100	$7.00
200	$14.00
500	$35.00

Phone Plan B

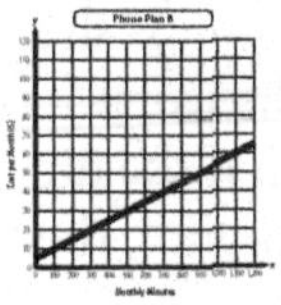

C = $5.00 + $0.05M

Tel-Mobile

$5.00 Monthly Fee

5 cents/minute

Monthly Minutes	Cost
50	$7.50
150	$12.50
250	$17.50
350	$22.50

Phone Plan C

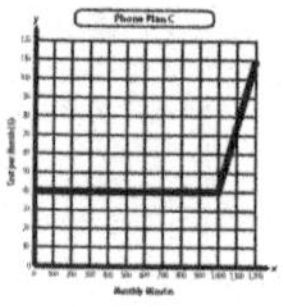

for $M \le 1{,}000$, $C = \$40$

for $M > 1{,}000$, $C = \$40 + \$0.35\,(M - 1{,}000)$

1,000 Anytime Minutes for $40.00/Month

$0.35 each additional minute

Monthly Minutes	Cost
500	$40.00
1,000	$40.00
1,100	$75.00
1,200	$110.00

Phone Plan D

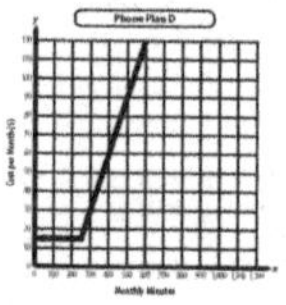

for $M \le 250$, $C = \$15$

for $M > 250$, $C = \$15 + \$0.30\,(M - 250)$

250 Anytime Minutes

$15 per month

(30 cents each additional minute)

Monthly Minutes	Cost
100	$15.00
150	$15.00
250	$15.00
300	$30.00

Activity 2: It Would Depend on the Person

Answers will vary. Sample answers:

1. Mary Jane should choose the NuWave plan, but has to be careful not to go over the 1,000 minutes. If she does go over, even by 30 minutes, Tel-Mobile would be better. The worst plan for her is Z-Phone because after the first 250 minutes, she would have to pay $0.30 a minute. If she talked only 2,000 minutes, that would mean she would have to pay $525 plus the $15 per month fee.
2. Jenny should choose the Nuwave plan because she could talk for up to 1,000 minutes without worrying about going over $40 a month. If she talks 30 minutes per night for a month, that equals about 900 minutes. The worst plan for her is Z-Phone because after the first 250 minutes, she would have to pay $0.30 a minute. If she talked 900 minutes, that would mean she would have to pay $195 plus the $15 per month fee.
3. Tricia should choose the Cellular plan because she would not have to pay anything unless she used her phone in an emergency. The worst plan for her is Nuwave because, even if she did not use her phone at all during the month, she would have to pay $40.
4. Answers will vary.

Practice: I Am Changing the Rules!

Answers will vary. Sample answers:

The new equation: $C = \$0.07M + \$20.00$

The new table:

Table 1

Monthly Minutes	Cost
50	$23.50
100	$27.00
200	$34.00
500	$55.00

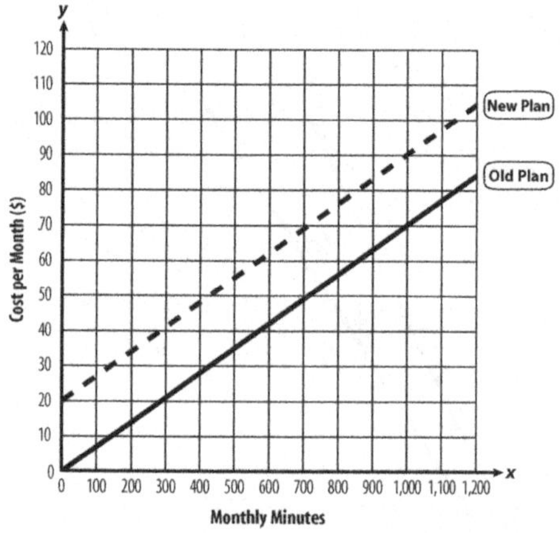

Symbol Sense Practice: Greater Than, Less Than

2. When $A \ge 65$, $C = \$0.05$.
   When $A < 65$, $C = \$1.25$.
3. When $A \ge 21$, $C = \$10.00$.
   When $A < 21$, $C = \$0.00$.
4. When $A \ge 5$, $C = \$250$.
   When $A < 5$, $C = \$125$.

Symbol Sense Practice: Solving Two-Step Equations

1. $x = 5$
2. $x = 8$
3. $x = 100$
4. $x = 5$
5. $x = 20$
6. $x = 10$
7. $x = 2$
8. $x = 7$
9. $x = 40$
10. $x = 800$

11. $2x + 3 = 7; x = 2$

12. $5x + 3 = 28; x = 5$

13. $1{,}000x + 250 = 6{,}250; x = 6$

14. $7x - 4 = 59; x = 9$

15. $x/2 - 10 = 41; x = 102$

Extension: Looking at Four Graphs

Answers will vary. Sample answer:

Juania probably talks between 200 and 300 minutes because that is the region where the graphs of A, *B*, and D are about equal ($15–$20). She said she would not pick Plan C because it would cost $40. Juania probably should look ahead and not take Plan D because it gets very expensive after the first 250 minutes. However, with Plan *B* the cost would increase more slowly when Juania goes over her usual usage.

Test Practice

1. (2)
2. (4)
3. (4)
4. (1)
5. (3)
6. 14.5

Lesson 10: Signs of Change

Activity 1: Seeing the Constant Rate of Change

Answers will vary. Sample answers:

1. The rate of change is steadily increasing by $15 per hour.
2.

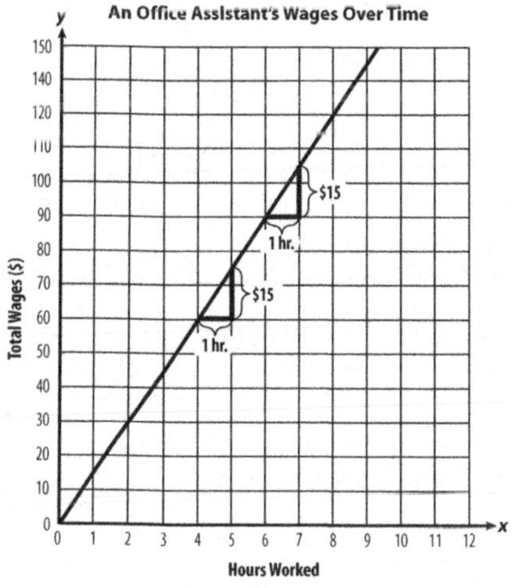

3. It is the same, or constant, because no matter what hour you look at, there is always a $15 increase from the hour before.

4. The rate of change at $18 per hour increases more quickly than the rate at $15 per hour. The $18 per hour slope is steeper.

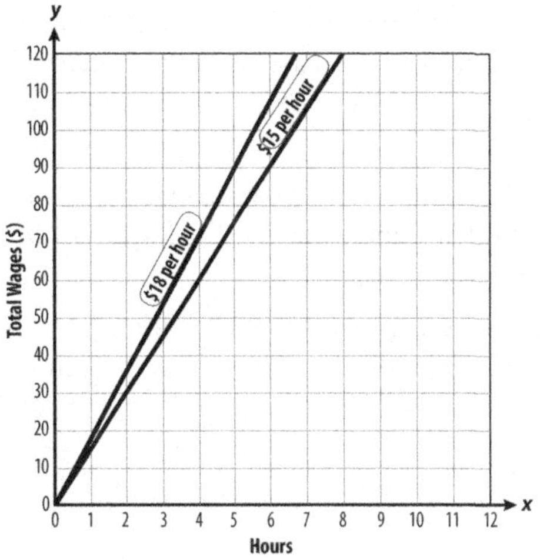

5. The rate of change would be the same for $15 per hour, but it would begin at point (0,100) instead of (0,0). The lines would be parallel, but total wages for the $18 per hour job would be higher on the grid at every point.
6. The more money earned per hour, the steeper the slope.

Activity 2: Ranking Rates of Change

1. Rankings of representations from slowest rate of change to fastest rate of change:
   (1) Table C ($y = x$)
   (2) Graph F ($y = 3x + 100$)
   (3) Equation A ($y = 4x + 50$)
   (4) Table D ($y = 5x + 50$) and Graphic E ($y = 5x + 25$)
   (5) Equation B ($y = 8x - 7$)
2. $y = 8x - 7$. Explanations will vary. Sample answer: I first looked at the number that x was multiplied by (coefficient). The larger the number, the greater the rate of change, so the equation $y = 8x - 7$ has the steepest line. Or, I drew the graphs for each and saw that the graph for $y = 8x - 7$ was the steepest with the greatest slope.
3. Answers will vary. Sample answer: Table D ($5x + 50$) could be someone opening a savings account with $50, then every week adding $5 to it.

Practice: Watching Money Grow in a Table

Answers will vary. Sample answers:

1. The Eastbank's rate of change is an increase of \$25 per quarter. The U.S. Savings' rate of change is an increase of \$50 per quarter. People's Bank increases at a rate of \$85 per quarter.
2. People's Bank would have the steepest slope; Eastbank's slope would be the flattest of the three lines.
3. They all slope upward, but at different rates. People's Bank has the steepest slope and Eastbank has the flattest slope.

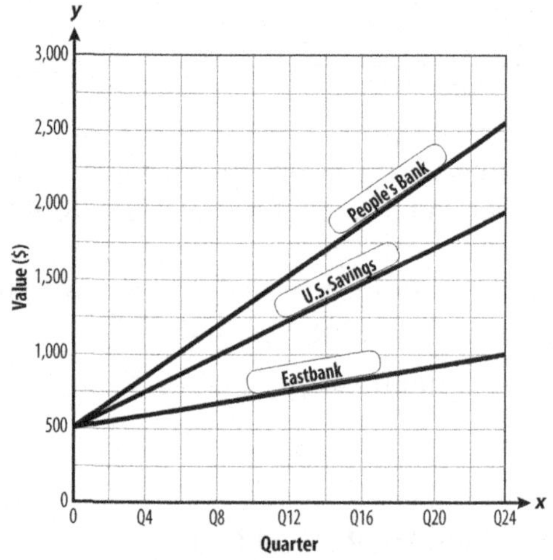

4. The three lines would be farther and farther apart from each other. The People's Bank slope would be the highest, with the most interest earned. Eastbank would show the least amount earned.
5. Eastbank's equation is $I = \$500 + \$25q$. U.S. Savings' equation is $I = \$500 + \$50q$. People's Bank's equation is $I = \$500 + \$85q$. The coefficients (multipliers of x) differ, but the addition of the constant amount of \$500 is the same for all three.

Practice: Whose Total Earnings Change the Fastest?

Answers will vary. Sample answers:

1. All the graphs are straight lines that slope upward. The assembler's rate of change is the flattest, so she earns less per hour than the other two job categories.
2. The mystery job might be a manager's job because that position earns \$30 an hour.
3.

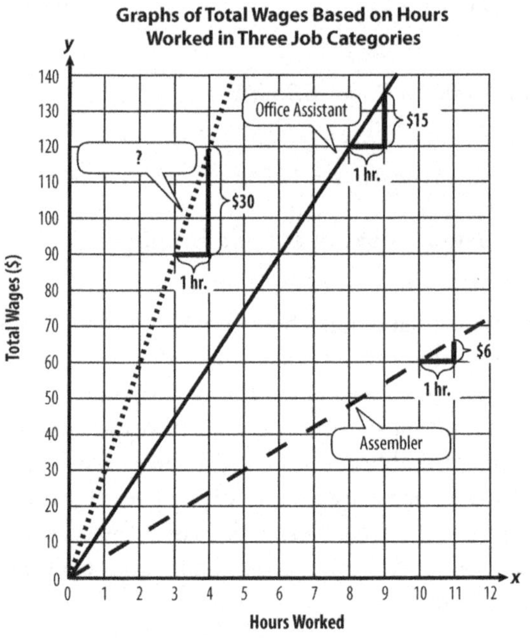

4. The mystery job position has the fastest constant rate of change. For every hour, it goes up \$15, whereas the other two jobs go up at rates of \$15 per hour and \$6 per hour.
5. The mystery job category's equation is $W = \$30h$. The office assistant's equation is $W = \$15h$. The assembler's equation is $W = \$6h$.
6. The three lines would be farther and farther apart from one another. The mystery job's slope would be the highest with the most wages earned over the year. The assembler's slope would be the flattest with the least amount of wages earned over the year.

Symbol Sense Practice: Solving More Two-Step Equations

1. $x = 7$
2. $x = 11$
3. $x = 23$
4. $x = 98$
5. $x = 15$
6. $x = 8$
7. $x = 24$
8. $x = 12$

Test Practice

1. (3)
2. (4)
3. (1)
4. (1)
5. (2)
6. 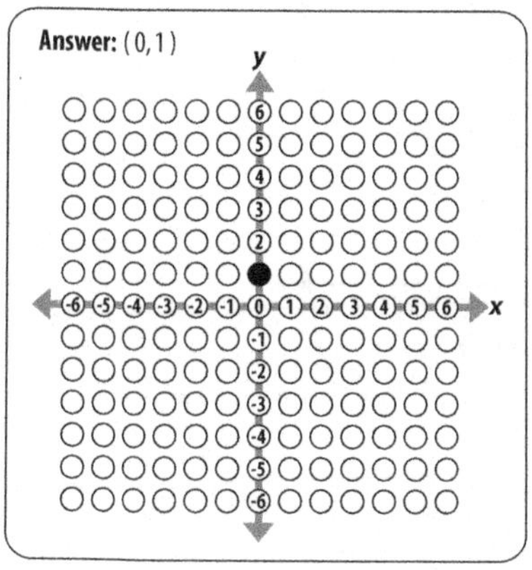

Lesson 11: Rising Gas Prices

Activity 1: Patterns of Change

1. The table and graph:

Case Number	Number of Dots Pattern 1	Number of Dots Pattern 2
1	5	1
2	9	4
3	13	9
4	17	16
5	21	25
10	41	100
25	101	625
n	$4n + 1$	n^2

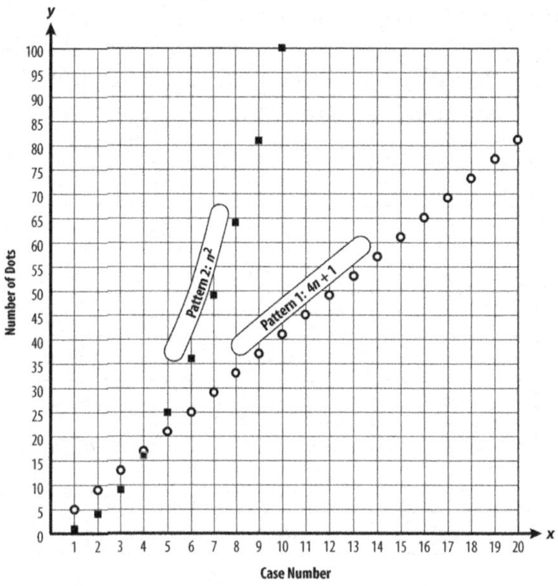

2. Sample explanation: The first pattern is linear. The pattern begins with 5, but constantly increases by 4 as the case number increases by 1. The second pattern is nonlinear; it increases by 3, 5, 7, 9, and so on.
3. In Pattern 1, the 10th entry will have 41 dots, the 25th entry will have 101 dots, and the nth entry will have $4n + 1$ (or $4(n - 1) + 5$) dots.

   In Pattern 2, the 10th entry will have 100 dots, the 25th entry will have 625 dots, and the nth entry will have n^2 dots.
4. Sample explanations: Pattern 2 grows faster because the number is always being multiplied by itself, so the bigger the number, the bigger the multiplier. In Pattern 1, the multiplier (coefficient) is always the same: 4. Or in Pattern 2, the amount added to the y-value gets successively larger, whereas in Pattern 1, the amount added to y in each successive case is 4, a constant rate of change.

Activity 2: Sky-High Gas Prices

1.

Month Number	Price per Gallon at Joe's	Price per Gallon at Metric
0	\$ 1.50	\$ 1.50
1	\$ 3.50	\$ 3.00
2	\$ 5.50	\$ 6.00
3	\$ 7.50	\$ 12.00
4	\$ 9.50	\$ 24.00
5	\$11.50	\$ 48.00
6	\$13.50	\$ 96.00
12	\$25.50	\$6,1440.00
n	$\$1.50 + 2n$	$\$1.50(2)^n$

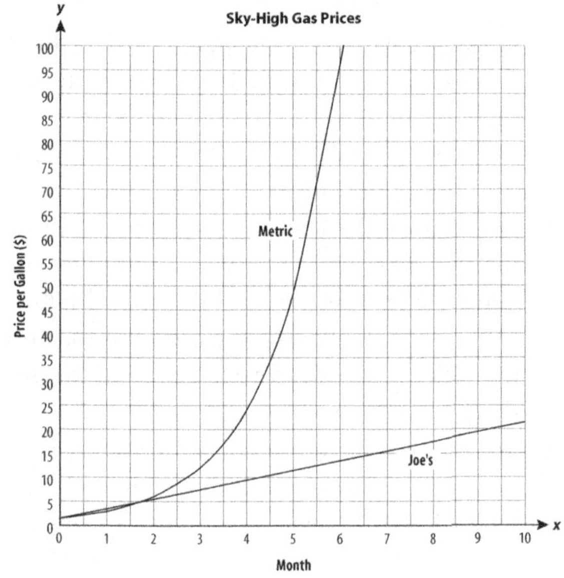

2. Joe's gas price is changing at a constant rate; each month it increases the same amount: \$2.00.
3. Sample explanation: Month 4 is when the cost per gallon at Metric jumps past Joe's and the gap between the two businesses starts to widen.
4. At the end of the year, Joe's gas price will be \$25.50 per gallon, and Metric's will be \$6,144 per gallon. Joe's will follow the formula $C = \$1.50 + 2n$, where C is the cost per gallon, and n is the number of months. After 12 months, $C = \$1.50 + \$2(12) = \$1.50 + \$24.00 = \$25.50$. Metric's will follow the formula $C = \$1.50(\$2)^n$. After 12 months, $C = \$1.50(\$2)^{12} = \$1.50(\$4{,}096) = \$6{,}144$.

Practice: More Patterns of Change

Patterns 1, 3, and 4 change at a constant rate.

Graphs will vary. Sample graph:

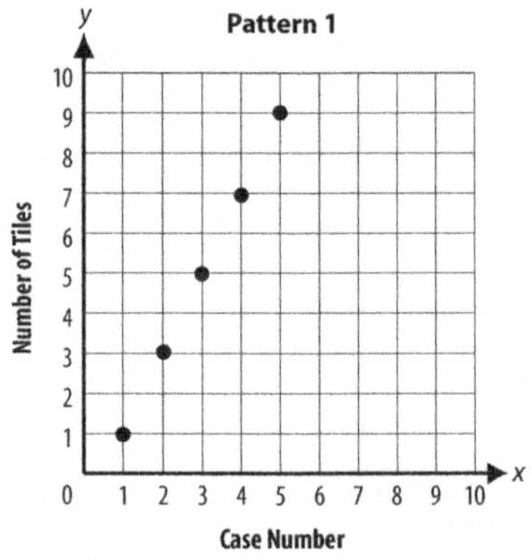

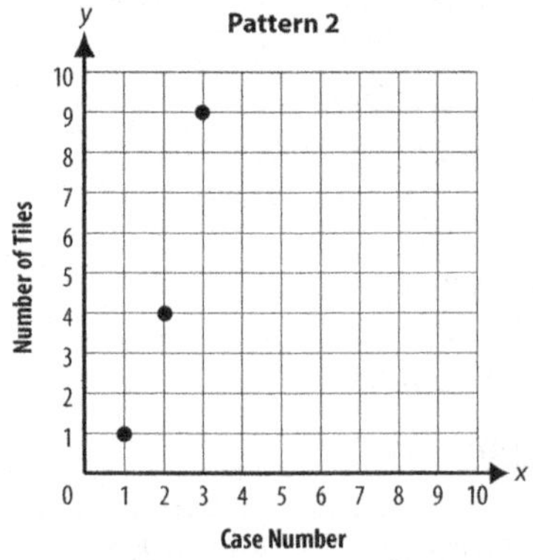

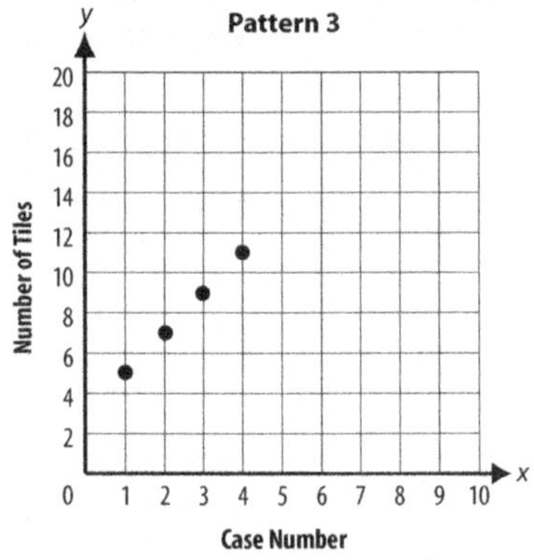

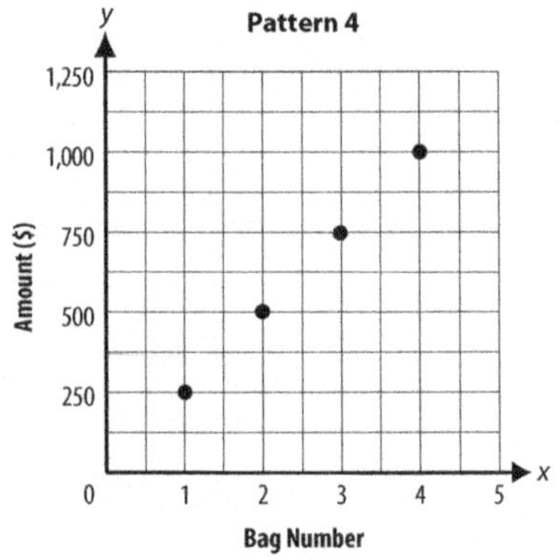

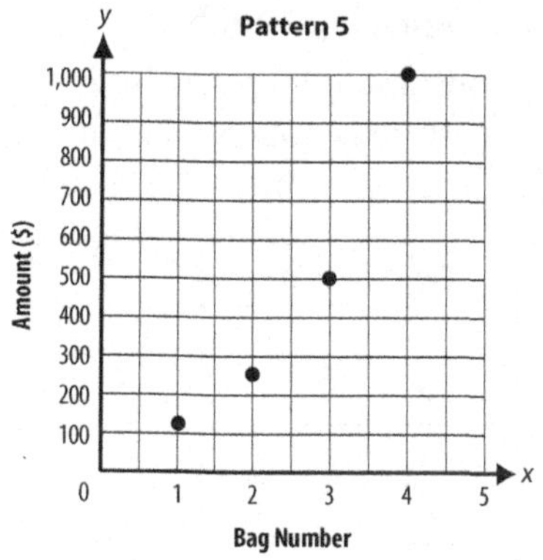

Practice: Investigation A—Expanding Squares

1. and 2.

Square	Perimeter	Area
1	4	1sq. unit
2	8	4 sq. unit
3	12	9 sq. unit
4	16	16 sq. unit
5	20	25 sq. unit
6	24	36 sq. unit
7	28	49 sq. unit
8	32	64 sq. unit

3. The perimeter = 60 units. The area = 15(15) or 225 square units.
4. When you lengthen the side of a square by one unit, the perimeter increases by four units. The area increases by $2n + 1$ (one more than twice the number of units on a side).
5. The rate of growth for the area is faster because you are multiplying the length of a side by itself. For the perimeter, you are always multiplying the length of a side by four.
6. The perimeter grows at a constant rate. It is a straight line that always increases by 4 as the side increases by 1.

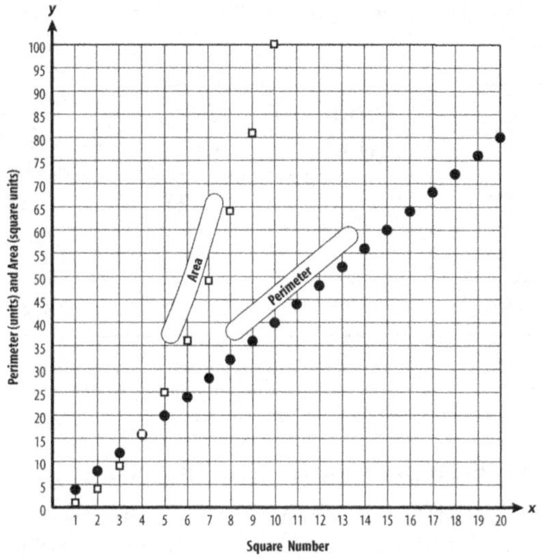

Practice: Investigation B—What Is Behind the Door?

1. Sample answer: I would pick the first door because the money keeps doubling from the day before, giving me a lot more money ($5,120) than the second door ($1,000). With door 2, I would just get $100 added each day, so I am always adding only 100.

2.

Day	Door 1	Door 2
1	$ 10	$ 100
2	$ 20	$ 200
3	$ 40	$ 300
4	$ 80	$ 400
5	$ 160	$ 500
10	$5,120	$1,000

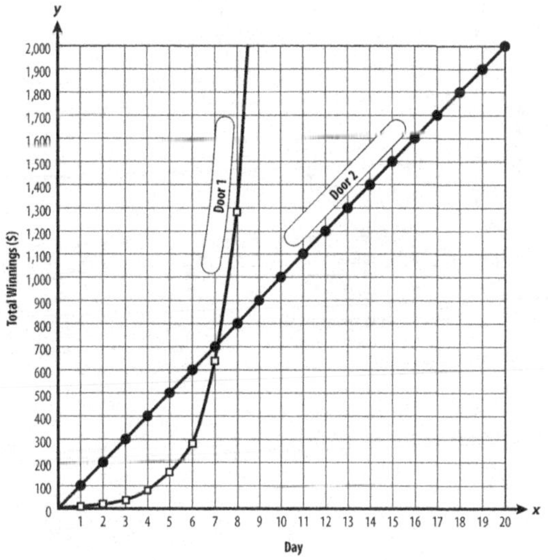

3. Door 2 grows at a constant rate because each day it always increases by exactly 100.

Practice: Investigation C—What Size Is That Copy?

%	Length (in.)	Width (in.)	Area (sq. in.)
50	3	1.25	3.75
100	6	2.5	15
150	9	3.75	33.75
200	12	5	60

1. Each time you increase the enlargement setting by 50%, the length increases by 3, the width increases by 1.25, but the area does not increase at a constant rate.
2. The length (and width) grows at a constant rate.

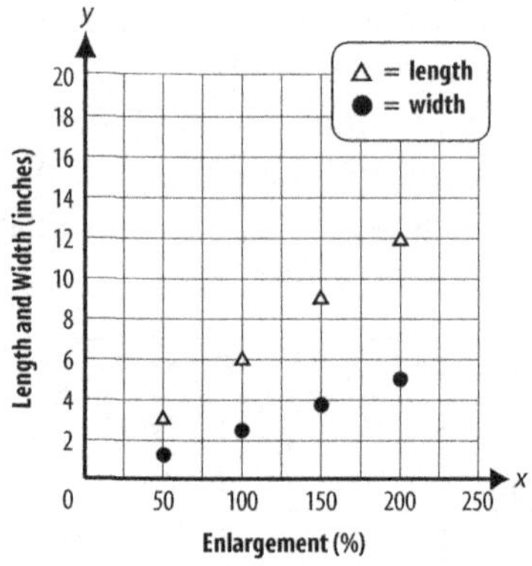

3. The area does not grow at a constant rate.

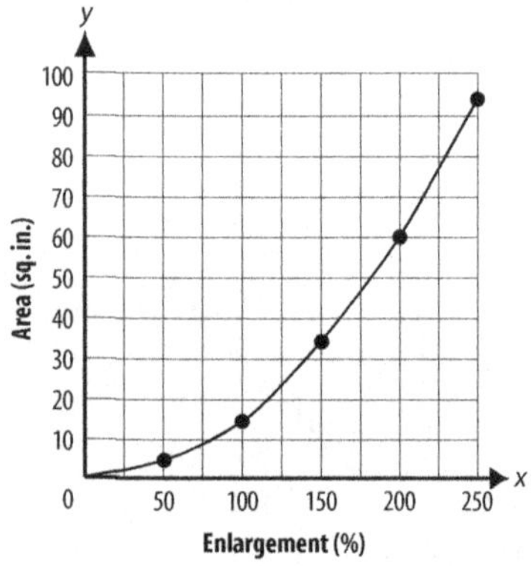

Symbol Sense Practice: Squaring Numbers

1. The $2n$ pattern is constant; it increases by 2 each time. The n^2 pattern is not constant; it increases by 3, 5, 7, 9 … each time.

n	$2n$	n^2
0	0	0
1	2	1
2	4	4
3	6	9
4	8	16
5	10	25
6	12	36
7	14	49
8	16	64
9	18	81
10	20	100
11	22	121
12	24	144

2.

n	$2n$	n^2
2.5	5	6.25
50	100	2,500
100	200	10,000
20	40	400

Symbol Sense Practice: Using Exponents

10^4 Ten to the fourth power
10 x 10 x 10 x 10
10,000

5^2 Five squared
5 x 5
25

2^5 Two to the fifth power
2 x 2 x 2 x 2 x 2
32

4^3 Four to the third power, or four cubed
4 x 4 x 4
64

3^4 Three to the fourth power
3 x 3 x 3 x 3
81

8^3 Eight to the third power, or eight cubed
8 x 8 x 8
512

10^3 Ten cubed, or ten to the third power

10 x 10 x 10

1,000

16^2 Sixteen squared

16 x 16

256

0^4 Zero to the fourth power

0 x 0 x 0 x 0

0

1^4 One to the fourth power

1 x 1 x 1 x 1

1

Test Practice

1. (4)
2. (4)
3. (3)
4. (1)
5. (3)
6. 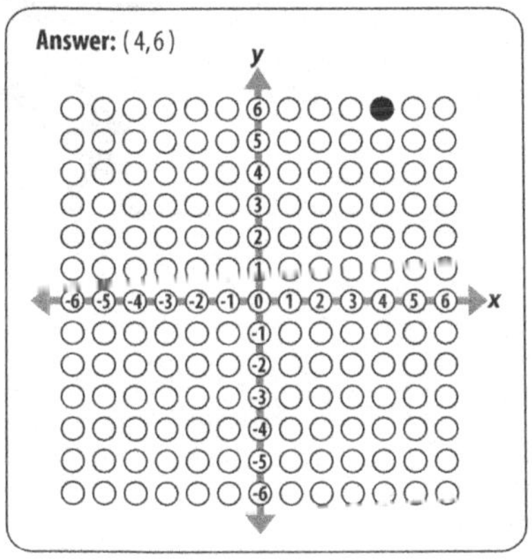

Lesson 12: The Patio Project

Activity: The Patio Project

Patio Size	Corner Tiles	Edge Tiles	Center Tiles	Total
2 by 2	4	0	0	4
3 by 3	4	4	1	9
4 by 4	4	8	4	16
5 by 5	4	12	9	25
6 by 6	4	16	16	36
100 by 100	4	392	9,604	10,000
n by n	4	$4(n-2)$	$(n-2)^2$	n^2

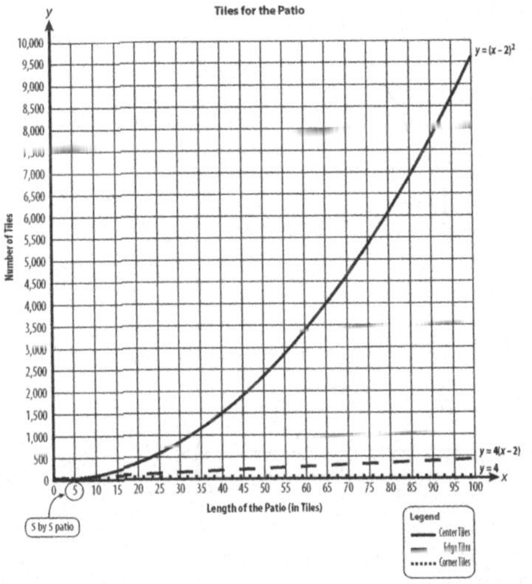

Closing the Unit: Pulling It All Together

Final Assessment

Task 1: Some Personal Patterns

Answers will depend on the situation chosen. Sample is for "Evalina drinks three cups of coffee a day."

Table:

Day	Cups of Coffee
1	3
2	6
3	9
5	15
7	21
10	30

Graph:

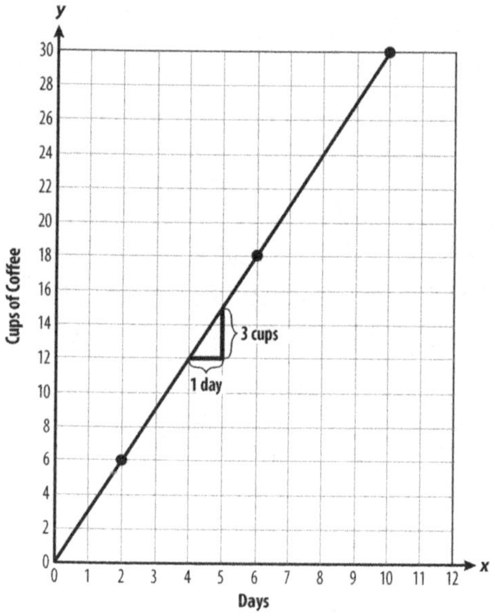

The rule in words: For any number of days, multiply the number of days by 3 to get the total number of cups of coffee Evalina drinks over time.

The equation: $y = 3x$, where y is the total cups of coffee and x is the number of days

Rate of change: For every day, the number of cups of coffee is always 3 more than the day before. In the equation, the rate of change is represented by the 3 in $3x$.

Task 2: Match Situations with Representations

Situation 1:

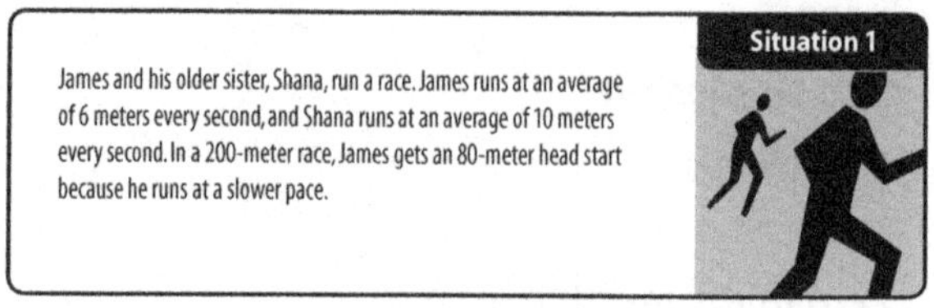

$y = 6x + 80$

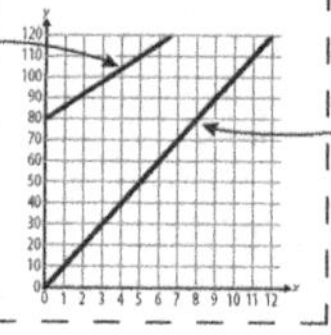

$y = 10x$

x	y
1	86
2	92
3	98
4	104

x	y
1	10
2	20
3	30
4	40

Linear. Constant rate of change for Shana ($y = 10x$) is 10 meters per second; constant rate of change for James ($y = 6x + 80$) is 6 meters per second.

Situation 2:

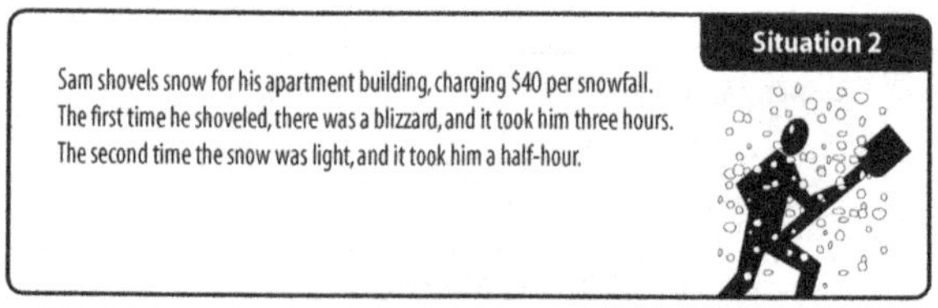

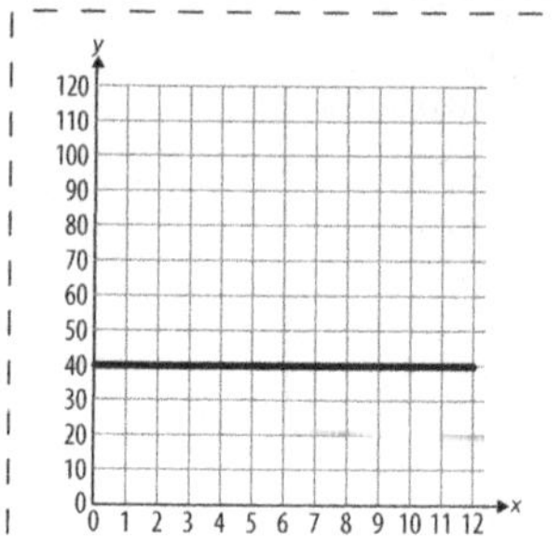

$y = 40$

x	y
1	40
2	40
3	40
4	40

Linear. Constant rate of change for Sam is 0.

Situation 3:

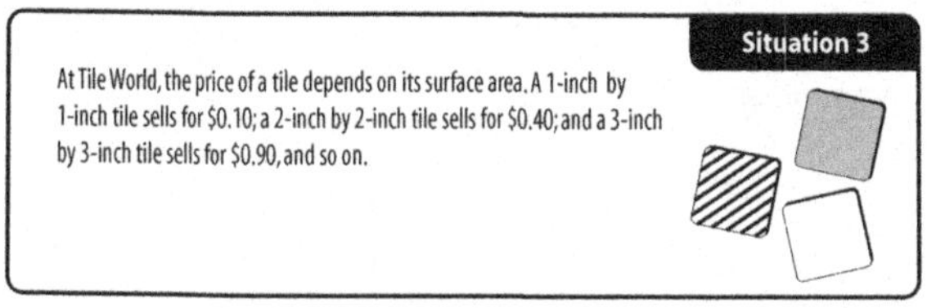

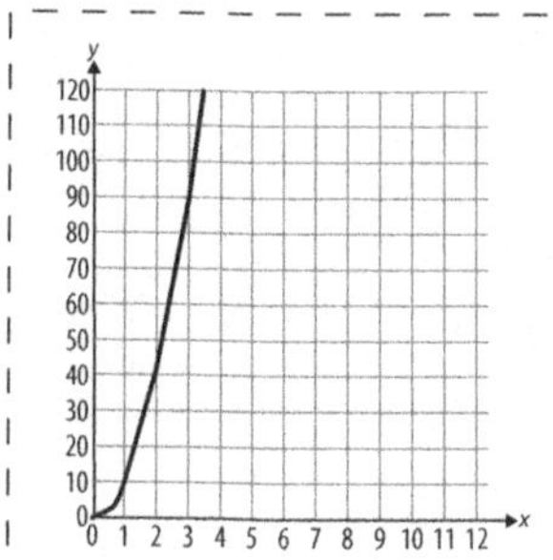

$y = 10x^2$

x	y
1	10
2	40
3	90
4	160

Nonlinear

Situation 4:

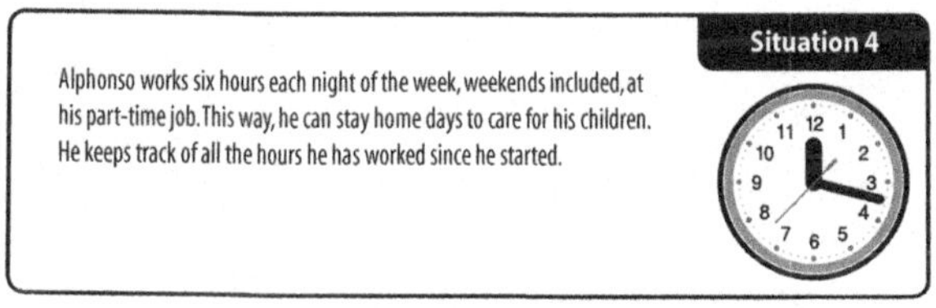

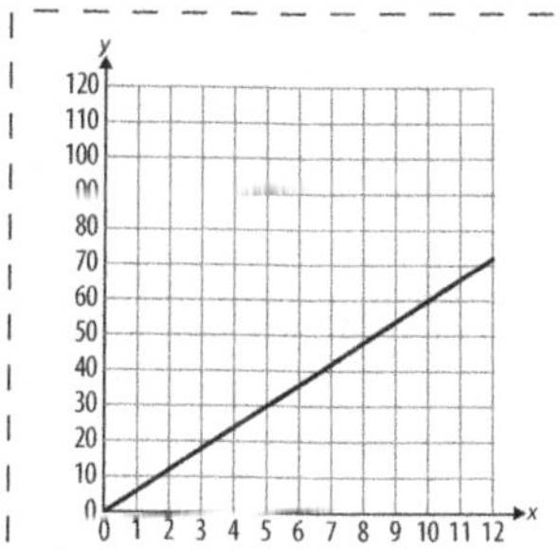

$y = 6x$

x	y
1	6
2	12
3	18
4	24

Linear. Constant rate of change for Alphonso ($y = 6x$) is $6 per hour.

Situation 5:

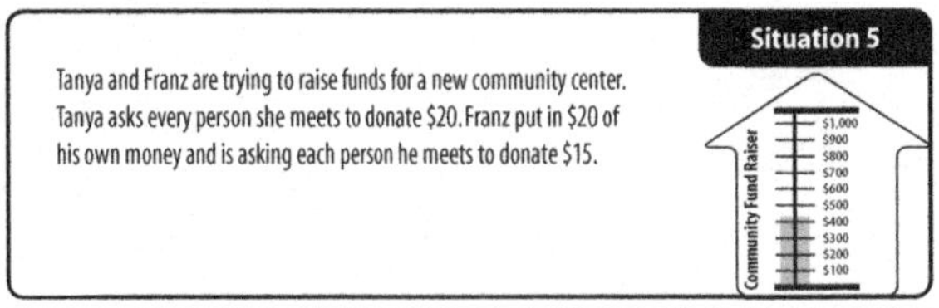

$y = 15x + 20$

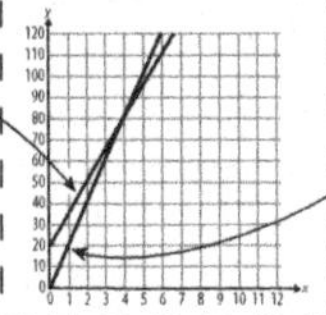

$y = 20x$

x	y
1	35
2	50
3	65
4	80

x	y
1	20
2	40
3	60
4	80

Linear. Constant rate of change for Tanya ($y = 20x$) is $20 per donation; constant rate of change for Franz ($y = 15x + 20$) is $15 per donation.

1. Franz will have $1,520. $y = 15x + 20$, so $y = 15(100) + 20 = 1{,}500 + 20 = 1{,}520$.
2. Never. The only time Franz and Tanya will have the same amount of collections is at $80.
3. It will be a tie. Each will take 20 seconds to run 200 meters. Shana; $y = 10x$, so 200 meters = 20(10). James: $y = 6x + 80$, so $200 = 6(20) + 80$.

Task 3: Symbol Sense

1. a. $6 + 3(12) = 42$
   b. $5(12)(8) = 480$
   c. $2(12) + 3(8) = 48$
   d. $(12 - 8)/2 = 2$
   e. $(12 + 8)(12 - 8) = 80$
   f. $(12)(8)^2 = 768$
   g. $[(12)(8)]^2 = 9{,}216$
   h. $(8)^2 (12)^2 = 9{,}216$
2. a. $x = 20$
   b. $x = 11$
   c. $x = 31$

Sources and Resources

Mathematics Education

The National Council of Teachers of Mathematics (NCTM) publishes several excellent resources.

- *Principles and Standards for School Mathematics.* Reston, VA: NCTM, 2000.
- The NCTM journals: *Mathematics Teaching in the Middle Grades, Mathematics Teacher*, and *Journal for Research in Mathematics Education.*
- http://www.nctm.org.

Fendel, D., D. Resek, L. Alper, & S. Fraser. *Interactive Mathematics Program.* Berkeley, CA: Key Curriculum Press, 1997.

Lappan, G., J. Fey, W. Fitzgerald, S. Friel, & E. Phillips. *Connected Mathematics Series.* Parsippany, NJ: Dale Seymour Publications, Division of Pearson Education, 1998.

National Research Council. *Adding It Up: Helping Children Learn Mathematics.* J. Kilpatrick, J. Swafford, and B. Findell (Eds.). Mathematics Learning Study Committee, Center for Education, Division of Behavioral and Social Sciences and Education. Washington, DC: National Academy Press, 2001.

Russell, S.J., et al. *The Investigations in Number, Data, and Space Curriculum.* White Plains, NY: Dale Seymour Publications, 1998.

Mathematics and Numeracy Education for Adults

Curry, D., M.J. Schmitt, and S. Waldron. *A Framework for Adult Numeracy Standards: The Mathematical Skills and Abilities Adults Need to Be Equipped for the Future.* Boston: World Education, 1996. (This framework was developed by members of the Adult Numeracy Network.) https://adultnumeracynetwork.org

Massachusetts Department of Education. *Massachusetts Adult Basic Education Curriculum Frameworks for Mathematics and Numeracy.* Malden, MA: Massachusetts Department of Education, 2001. https://www.doe.mass.edu/acls/frameworks.

Stein, S. *Equipped for the Future Content Standards: What Adults Need to Know and Be Able to Do in the Twenty-First Century.* ED Pubs document EX0099P. Washington, DC: National Institute for Literacy, 2000.

Algebra

Cramer, K. "Using Models to Build an Understanding of Functions." *Mathematics Teaching in the Middle School* 6, no. 5 (Jan. 2001): 310–18.

Driscoll, M. *Fostering Algebraic Thinking: A Guide for Teachers in Grades 6 Through 10.* Portsmouth, NH: Heinemann, 1999.

King, J., & P. Rasmussen. *Key to Algebra* series. Emeryville, CA: Key Curriculum Press, 1990.

Lawrence, A., & C. Hennessy. *Lessons for Algebraic Thinking: Grades 6–8.* Sausalito, CA: Math Solutions Publications, 2002.

Mason, J. "Expressing Generality and Roots of Algebra." In Bednarz, N., C. Kieran, and L. Lee (eds.), *Approaches to Algebra: Perspectives for Learning and Teaching*, 65–86. Dordrecht, The Netherlands: Kluwer Academic Publishers, 1996.

Moses, B. (ed.). *Algebraic Thinking: Grades K–12, Readings from NCTM's School-Based Journals and Other Publications.* Reston, VA: NCTM, 1999.

National Council of Teachers of Mathematics. *The Ideas of Algebra, K–12.* Reston, VA; NCTM, 1988.

National Council of Teachers of Mathematics. *Navigating Through Algebra in Grades 6–8.* Reston, VA: NCTM, 2001.

National Council of Teachers of Mathematics. *The Roles of Representation in School Mathematics.* Reston, VA: NCTM, 2001.

National Research Council. *The Nature and Role of Algebra in the K–14 Curriculum: Proceedings of a National Symposium. May 28 & 29,* 1997. Washington, DC: National Academy Press, 1998.

There are some very thought-provoking papers in this book, especially Jim Kaput's "Transforming Algebra from an Engine of Inequity to an Engine of Mathematical Power by 'Algebrafying' the K–12 Curriculum."

Radford, L. "Signs and Meanings in Students' Emergent Algebraic Thinking: A Semiotic Analysis." *Educational Studies in Mathematics* 42, no. 3, (2000): 237–68.

Steen, L. *On the Shoulders of Giants: A New Approach to Numeracy.* Mathematics Sciences Education Board and the National Research Council. Washington, DC: National Academy Press, 1990.

Web Sites

For great interactive Java applets for students or teachers see
http://www.shodor.org/interactivate/activities/tools.html

For a great course for teachers, see Patterns, Functions, and Algebra,
http://www.learner.org/channel/courses/learningmath/algebra

For math questions see "Ask Dr. Math®"
https://www.nctm.org/drmath/

Made in United States
North Haven, CT
05 November 2026

21778839R00161